Student Traitbook

Vicki Spandel and Jeff Hicks

www.greatsource.com
800-289-4490

Acknowledgements

For permission to reprint copyrighted material, grateful acknowledgement is made to the following sources:

Excerpt from *After Hamelin* by Bill Richardson. Text copyright © 2000 by Bill Richardson. Reprinted by permission of Annick Press.

Excerpt from *All Quiet on the Western Front* by Erich Maria Remarque. Text copyright © 1929, 1930 by Little, Brown and Company. Copyright renewed 1957, 1958 by Erich Maria Remarque. Reprinted by permission of New York University Press.

Excerpt from *Animal Farm* by George Orwell. Text copyright © 1946 by Harcourt Inc. Copyright renewed 1974 by Sonia Orwell. Reprinted by permission of Houghton Mifflin Harcourt Publishing Company and A.M. Heath & Co.

Excerpt from *Artemis Fowl* by Eoin Colfer. Text copyright © 2001 by Eoin Colfer. Reprinted by permission of Hyperion Books for Children and Penguin Group Ltd.

Excerpt from *Endurance* by Frank Arthur Worsley. Text copyright © 1931 by F.A. Worsley.

Excerpt from *The Endurance: Shackleton's Legendary Antarctic Expedition* by Caroline Alexander. Text copyright © 1998 by Caroline Alexander. Reprinted by permission of Alfred A. Knopf, Inc, a division of Random House, Inc.

Excerpt from *Exploding Ants: Amazing Facts About How Animals Adapt* by Joanne Settel, Ph.D. Text copyright © 1999 by Joanne Settel. Reprinted by permission of Atheneum Books for Young Readers, an imprint of Simon & Schuster Children's Publishing Division and Columbia Literary Associates, on behalf of the author.

Excerpt from *The Fellowship of the Ring* by J.R.R. Tolkien. Text copyright © 1954, 1965, 1966 by J.R.R. Tolkien. Copyright renewed in 1982, 1993, 1994 by Christopher R. Tolkien, John F.R. Tolkien and Priscilla M.A.R. Tolkien. Reprinted by permission of Houghton Mifflin Harcourt Publishing Company and HarperCollins Publishers Ltd.

Excerpt from *The Hungry Ocean: A Swordboat Captain's Journey* by Linda Greenlaw. Text copyright © 1999 by Linda Greenlaw. Reprinted by permission of Hyperion Books for Children and Hodder and Stoughton Limited.

Excerpt from *The Killer Angels* by Michael Shaara. Text copyright © 1974 by Michael Shaara. Copyright renewed by Jeff M. Shaara and Lila E. Shaara. Reprinted by permssion of Ballantine Books, a division of Random House, Inc.

Excerpt from *Red Scarf Girl: A Memoir of the Cultural Revolution* by Ji-Li Jiang. Text copyright © 1997 by Ji-Li Jiang.. Reprinted by permission of HarperCollins Publishers.

Excerpt from *The Secret Knowledge of Water* by Craig Childs. Text copyright © 2000 By Craig Childs. Reprinted by permission of Sasquatch Books and The Spieler Agency, on behalf of the author.

Excerpt from *Troy* by Adéle Geras. Text copyright © 2000 by Adéle Geras. Reprinted by permission of Houghton Mifflin Harcourt Publishing Company and Scholastic Inc.

Excerpt from *Undaunted Courage: Meriwether Lewis, Thomas Jefferson and the Opening of the American West* by Stephen E. Ambrose. Text copyright © 1996 by Ambrose-Tubbs, Inc. Reprinted by permission of Simon & Schuster Inc. and Ambrose & Ambrose, Inc.

Copyright © 2010 by Houghton Mifflin Harcourt Publishing Company

All rights reserved. No part of this work may be reproduced or transmitted in any form or by any means, electronic or mechanical, including photocopying or recording, or by any information storage or retrieval system, without the prior written permission of the copyright owner unless such copying is expressly permitted by federal copyright law.

Permission is hereby granted to individuals using the Write Traits program to photocopy entire pages from this publication in classroom quantities for instructional use and not for resale. Requests for information on other matters regarding duplication of this work should be addressed to Houghton Mifflin Harcourt Publishing Company, Attn: Paralegal, 9400 South Park Center Loop, Orlando, Florida 32819.

Printed in the U.S.A.

ISBN-13   978-0-669-01518-8

ISBN-10   0-669-01518-0

1 2 3 4 5 6 7 8 9 10   1409   18 17 16 15 14 13 12 11 10

4500237787

If you have received these materials as examination copies free of charge, Houghton Mifflin Harcourt Publishing Company retains title to the materials and they may not be resold. Resale of examination copies is strictly prohibited.

Possession of this publication in print format does not entitle users to convert this publication, or any portion of it, into electronic format.

About the Authors

Jeff Hicks

Jeff taught for 18 years in the Beaverton School District (home of the 6-traits) where he enjoyed working with students to help them find their voices as writers. He is the co-author of *Write Traits Classroom Kits*, *Write Traits Advanced*, and *Write Traits Kindergarten*. Though his heart is still in the classroom, he is now a full-time writer, presenter, and professional development consultant. He lives in Beaverton with his wife and son, and he currently serves on the Beaverton School Board.

Vicki Spandel

Vicki is a founding coordinator of the 17-member teacher team that developed the original, nationally recognized 6-trait model for writing assessment and instruction. A specialist in teaching writing and revision to students of all ages, she is the author of *Write Traits Classroom Kits*, *Write Traits Advanced*, and *Write Traits Kindergarten*, as well as *The 9 Rights of Every Writer*, *Creating Writers*, and *Creating Young Writers*. She makes her home in the town of Sisters, Oregon, bordering the beautiful Three Sisters Wilderness.

Contents

UNIT 1: IDEAS
1.1 Bringing Your Message Down to Earth................7
1.2 Thinking like a Reader..................... 13
1.3 Defogging with Questions 19
1.4 Freedom from Filler......................... 24

Conventions and Presentation:
- Priority Access.. 31
- Keep It Under Cover 37

UNIT 2: ORGANIZATION
2.1 Finding the Right Design................... 54
2.2 Putting Ideas First 61
2.3 Holding It Together 65
2.4 The Whole Package 71

Conventions and Presentation:
- Readers Need a Comma Break 76
- A Recipe for Clarity 81

UNIT 3: VOICE
3.1 A Defining Moment........................... 97
3.2 Writing with Confidence................. 103
3.3 Keeping Your Audience in Mind 109
3.4 Kicking It into High Gear 114

Conventions and Presentation:
- Real Speech 119
- Personalized Greetings....................... 124

UNIT 4: WORD CHOICE

4.1 The Right Shade of Meaning............ 142
4.2 Using All Your Senses 149
4.3 Getting Precise................................. 154
4.4 Stop the Clutter! 162

Conventions and Presentation:
- All Is Well and Good 167
- Words that Sell 172

UNIT 5: SENTENCE FLUENCY

5.1 Don't Repeat Unless You Mean It! 190
5.2 The Logical Flow of Ideas 197
5.3 Secrets to Fluency 204
5.4 Smooth Sailing................................. 210

Conventions and Presentation:
- Discriminating Fragments 215
- No Rules Poetry 220

UNIT 1
Ideas

Two things make the trait of **Ideas** work: a strong central message and all the details that bring that message to life. All writing begins with something to say—a story to tell, a concept to explain, an argument to make. That's why the trait of Ideas is foundational. Everything else—organization, voice, the words you choose, the way you form your sentences—rotates around your ideas to influence your message. And, the effectiveness of your message depends on finding the right details—just enough details—to grab and hold your readers' attention.

In this unit, you will practice strategies for doing just that. You'll learn how to

- narrow your topic.
- think like a reader.
- clear the "fog" by answering readers' questions.
- cut filler that may garble your message.

Ideas

Name _____ Date _____

Sample Paper 1

Score for Ideas _____

If I Were Stranded

You've probably heard of people making "deserted island" lists—what music, gear, tools, or whatever, would you want to have with you if you were stranded on a deserted island? So, here's a version of that question for you. What would you do if you were stranded on a deserted island (no people, no civilization), and you could take only <u>one</u> thing with you? Think carefully, now, before you just blurt out an answer.

I'll help you out by sharing some answers from my friends—a watch, a cell phone loaded with apps and games and music, a survival knife, a book, and instant pudding. (No joke. I have a friend who loves pudding.) Let's analyze these, starting with the ones that don't make much sense. Now, why would you want a watch? If you were stranded for any length of time, hours and minutes would cease to have meaning. A cell phone is also a pretty poor choice. If you had coverage, which you probably wouldn't, you could call someone and let them know you were stranded. But what are the chances your battery would last long enough to try to make calls, much less play with your apps or listen to music? Your phone would be a distraction, and then a frustration. You'd be better off spending your time looking for food and water. As to the pudding, what can we say? Really, man. *Pudding?*

Bringing a knife or a book makes more sense. The benefits of a knife are obvious—protection, a tool for making fire, shelter, or making the ends of sticks into spears. These are all good things. A book is a sensible answer, too. At least you would have entertainment and company. Of course, reading the same book over and over could make you crazy (unless it was a survival guide for people stuck on islands). You might be better off writing your own book, once you figured out what to write on and write with.

Ideas

Now, to my personal recommendation—I would take my fly-fishing rod. It's that simple. A person has to eat, and if you're on an island, it stands to reason there are fish around. I would keep myself entertained and occupied doing something I love to do. Being stranded on an island with no distractions or competition is a dream for someone who loves to fish, as I do—as long as I'm not stranded for too long. I'd be pretty happy for a month. Make that a year.

In short, a fly-fishing rod is the perfect answer–or at least perfect for me. Being stranded is about surviving, and usually we think that means getting food, water, and shelter, the basic needs. But you also need to help your mind survive. However you answer the question, make sure your choice helps you survive without losing the most important thing you'll take with you—a sound mind.

Ideas

Sample Paper 2

Score for Ideas _____

Moths

Have you ever seen a moth up close? It is one of the most beautiful things you will ever see. Moths have amazing colors, unlike any colors that you will see elsewhere. You might see shades of blue or green, gold or brown, depending on the kind of moth you are seeing. Some moths look furry or fuzzy when you see them at close range. Their antennae are totally amazing. They use them for many things. Moths are special creatures.

Moth wings have a lot of unusual patterns that show up as you look at them through a magnifying glass or microscope. These patterns are not random. They are actually very useful for the moth's survival. Butterflies are also very beautiful and have amazing patterns of colors and shapes on their wings, too. Butterflies and moths are similar in many ways.

Some people photograph or paint moths because they are like flying works of art.

Of course, some people do not like moths at all and will kill them if they can. People will spray them with chemicals or install a zapper to attract and kill them. Cats kill moths, as well. Moths can do a lot of damage to crops, especially when they are in the caterpillar stage. They lay eggs that hatch to become destructive caterpillars.

Moths live only about two or three weeks at the most. Butterflies and moths are really very beautiful, and we should enjoy them during their short lives.

Ideas

The WRITER... makes everything crystal clear, beginning to end.

So the READER...
_____
_____
_____

The WRITER... keeps the writing small and focused.

So the READER...
_____
_____
_____

The WRITER... chooses details with care.

So the READER...
_____
_____
_____

The WRITER... knows the topic inside and out.

So the READER...
_____
_____
_____

6 Unit 1

Ideas

Lesson 1.1

Bringing Your Message Down to Earth

Have you ever seen one of those pictures of the Earth taken from a satellite or space shuttle? Even though you might recognize Earth and say "Hey—that's home!", you would have to zoom closer and closer to see your continent . . . country . . . state or province . . . city . . . street . . . and finally your actual residence. In the early stages of writing, topics sometimes start out as big as that far-away view of Earth. Writing about such topics is a daunting task—and what's more, it results in generalizations: *Earth is a special place.* Readers don't want generalizations. They want close-up, specific details: *Our place is home to five kids, three cats, an iguana named Harry, and a boa constrictor named Ethel.*

Hear the difference? In this lesson, you'll have a chance to take a daunting topic down to manageable size—the size you could write about in 15 minutes.

Sharing an Example: *Exploding Ants*

In her book, *Exploding Ants: Amazing Facts About How Animals Adapt,* author and biology professor Joanne Settel, Ph.D., focuses on the ingenious ways some creatures find to survive. This precise focus actually makes her writing easier since there are so many things she doesn't need to include. Consider this sample passage about how the liver fluke survives by getting itself eaten—not just once, but three times.

Unit 1 • Lesson 1.1   7

Ideas

Name _____ Date _____

    The tiny wormlike fluke is a parasite that spends different parts of its life inside the bodies of three different host animals: a snail, an ant, and a sheep. The fluke must get inside each host by being eaten. It uses its amazing reprogramming skills to get itself into the mouth of a hungry sheep.

    Liver flukes actually begin their lives inside a snail. The snail starts things off when it eats some sheep dung filled with liver fluke eggs. Inside the snail, the eggs hatch, releasing thousands of tiny larvae (young flukes). As many as six thousand mucus-covered fluke larvae then gather together into a squirming ball. Eventually the snail ejects this grape-sized glob from its body.

    The next step in the fluke's life cycle takes place when an ant feeds on the mucus ball. This brings the larvae into the ant's stomach, where they bore into the stomach wall. Most of the larvae remain here and grow into adult flukes.

Exploding Ants
by Joanne Settel, Ph.D.

What Did You Learn?

Write three things you learned about flukes from the close-up view in this passage:

1. _____

2. _____

3. _____

Which of the following comes closest to stating the main idea of Settel's passage?

☐ The world is full of interesting animals.

☐ The tiny fluke is a parasite with an unusual life cycle.

☐ Sheep actually survive by eating flukes.

Ideas

Test the Doctor

Take one more look at Dr. Settel's passage on liver flukes, pencil in hand. As you go through it, mark each sentence **G** for general (*Liver flukes are cool!*) or **S** for specific—meaning the statement teaches you something about liver flukes. How did Dr. Settel do?

From Galactic to Specific

We don't know exactly how Dr. Settel came up with the idea to write about liver flukes, but let's imagine she started with a gigantic topic: *Living Things*. She would certainly have had to narrow that topic down—and it might have taken her several steps to get from *Living Things* to *Liver Flukes*. Her steps might have looked something like this:

- Living things (Galactic)
- Animals
- Animals with unusual life cycles
- Parasites
- Parasites with life cycles connected to ants
- Liver flukes (Specific)

Now imagine a writer creating another essay. Following are some topics the writer might consider. They're listed randomly. Work with a partner to put them in order—from galactic to specific. As you work, imagine yourself as a kind of human search engine, bringing in the focus tighter and tighter, until you bring your topic from outer space . . . right down to Earth.

Random List of Topics

- The Venus flytrap
- Plants
- The difficulties of growing a Venus flytrap as a potted plant
- Living things
- Unusual plants
- Carnivorous plants

Unit 1 • Lesson 1.1

Ideas

Name _____ Date _____

Same Topics Arranged From Galactic to Specific

1. _____
2. _____
3. _____
4. _____
5. _____
6. _____

Reflecting

Reflect for a moment on this narrowing process. How does a writer know when a topic is small enough to write about?

_____
_____
_____

Narrowing Your Own Topic

Let's suppose your teacher has asked you to write a short essay on one of the following topics:

- Ethics
- Sports
- Politics
- Entertainment
- Education

You'd need a lifetime to research any one of these topics. Unfortunately, you only have about 20 minutes to prewrite (narrow your topic) and write a short essay. So—this is about survival!

Ideas

Name _____ Date _____

Start by narrowing your topic. It may take you two or three steps—or more. Work with a partner. Begin with the same BIG topic; you do not have to wind up with identical narrow topics, however. Once you reach a certain point in the narrowing process, your personal interests may pull you in different directions.

Our BIG topic: _____

Going smaller and smaller . . .

1. _____
2. _____
3. _____
4. _____
5. _____

My focused, narrow topic: _____
_____

Small Topic = Short Essay

If you've narrowed your topic enough, this part should be easy! On your own scratch paper, write for 15 minutes, sharing what you know about your small topic. Make it personal and focused. Think *my home* versus *Earth from outer space*. Keep the pencil moving—and keep your reader's attention.

Share and Compare

Share your essay with a partner or in a writing circle. Did everyone wind up with a topic small enough to manage easily? If not, talk about ways to narrow some topics further.

Ideas

Name _____ Date _____

A Writer's Question
What happens to the writer's details when the topic is too big?

Putting It to the Test

How could this practice of narrowing a topic help you in a testing situation? For example, what could you do with these common writing prompts?

- A Day I'll Always Remember
- A Memorable Experience
- The Most Important Invention Ever

Ideas

Name .... Date ....

Lesson 1.2

Thinking like a Reader

Every once in a while, a great writing idea pops into your head, already narrowed down and crystal clear. When that happens, reach for the nearest pencil or keyboard! If your idea is still a little fuzzy, however, you might reach instead for something called a T-table. A T-table is a graphic organizer that can help writers *and* readers focus on thoughts and feelings. In this lesson, you'll start out as a reader, using a T-table to record your responses. Then, as a writer, you'll use the same graphic to capture the thoughts and feelings you hope *your* readers will have.

A Nonfiction Example: *Undaunted Courage*

This example of informational writing comes from Stephen E. Ambrose's book *Undaunted Courage: Meriwether Lewis, Thomas Jefferson, and the Opening of the American West.* In this passage, Ambrose talks about startling changes that occurred in America between 1800 and 1860. Read the passage. Then use the T-table that follows to record what you **see** and **feel** as you read. (Just write notes, not whole sentences.)

Since the birth of civilization, there had been almost no changes in commerce or transportation. Americans lived in a free and democratic society, the first in the world since ancient Greece, a society that read Shakespeare and had produced George Washington and Thomas Jefferson, but a society whose technology was barely advanced over that of the Greeks. The Americans of 1801 had more gadgets, better weapons, a superior knowledge of geography, and other advantages over

Ideas

Name _____ Date _____

the ancients, but they could not move goods or themselves or information by land or water any faster than had the Greeks and Romans...

But only sixty years later, when Abraham Lincoln took the Oath of Office as the sixteenth president of the United States, Americans could move bulky items in great quantity farther in an hour than Americans of 1801 could do in a day, whether by land (twenty-five miles per hour on railroads) or water (ten miles an hour upstream on a steamboat). This great leap forward in transportation—a factor of twenty or more—in so short a space of time must be reckoned as the greatest and most unexpected revolution of all—except for another technological revolution, the transmitting of information. In Jefferson's day, it took six weeks to move information from the Mississippi River to Washington, D.C. In Lincoln's, information moved over the same route by telegraph all but instantaneously.

Undaunted Courage
by Stephen E. Ambrose

Ideas

Name _____ **Date** _____

Reflection

Do you think that author Stephen E. Ambrose planned or hoped for you to see and feel certain things? Share your thoughts here.

_____
_____
_____

Another Example: "A Toast to Traditions"

This writing sample comes from a student writer. Read it carefully. Then use the T-table for this piece to record your thoughts and feelings. Since this is a personal narrative, what you feel or see in your mind as you read may be very different from your response to the Ambrose piece.

A Toast to Traditions

My family is very big on traditions. Thanksgiving, always a big holiday in our house, is loaded with them.

For one thing, dinner is always at my grandparents' house. It's very small and crowded for 20 of us, but squeezing into a tiny space is part of the tradition. We have a family toast where we clink glasses and all say, "*Zum Bahnhof!*" (which means something like "to the train station" in German). My grandmother loved to say this as a little girl—and the saying lives on.

We always have turkey—and part of our tradition is for Grandpa to complain about the "round bottomed" platter as he carves. The turkey wobbles back and forth precariously—but he won't use anything else.

One tradition I could forego is the vegetable aspic salad. (Imagine wiggly, brownish gelatin with vegetables and olives—every kid's favorite.) It looks and tastes disgusting, but everyone has to take one bite. *No excuses.*

Unit 1 • Lesson 1.2

Ideas

Name _____ Date _____

Traditions, like them or not, are strangely comforting—maybe because they're predictable. As much as I hate aspic salad, its lingering flavor is fixed in my mind as part of our Thanksgiving tradition.

Reflection

Do you think the author of this personal narrative planned for readers like you to see and feel certain things? Share your thoughts here.

_____
_____

Connecting to the Traits

Whether it's a personal narrative or an informational piece, some writing creates clear pictures or impressions in the reader's mind. The writer clearly knows the topic well. That's the trait of _____ in action.

With some writing, the reader senses a strong commitment to the topic. The writer cares deeply about the message and seems to want the reader to feel the same way. That's the trait of _____ in action.

Ideas

Name _____ Date _____

T-Off, In Reverse

You've been using a T-table to record your impressions as a reader. As a writer, you can also use it to plan. It's a way of getting in touch with your readers, imagining how you'd like them to think and feel as they read your work. Begin by choosing an idea. Write about anything that is on your mind right now, or use our list to help you think of a topic:

- Traditions I love—or would change if I could
- A memory linked to my grandparents
- Recent changes in technology
- Changes in transportation
- A food I could live without

The Two-Minute Conference

Meet with a partner and talk about your idea, either sharing information or telling your story. Take two minutes—no more. Then let your partner share with you, listening carefully for two minutes and asking questions.

T-table Planning

Using the T-table that follows, write notes to yourself about four to six things you really want/need readers to clearly *see* in their minds as they read your piece. Then describe the *feelings* you want your readers to have about your topic as they read. Remember—no sentences, just short notes.

Ideas

Name _____ Date _____

What I Want Readers to See	What I Want Readers to Think or Feel

From Planning to Drafting

Use your own paper to begin this piece of writing. Go as far as you can in about 15 minutes. Remember to

- focus on what you want the reader to *see*.
- think about how you want the reader to *feel*.
- look back at your T-table if you get stuck.

Hint: Think like a reader. Write what you would want to read.

A Writer's Questions
Could a T-table also be useful in guiding revision?

Putting It to the Test
Would a T-table be a good prewriting strategy to use in a writing assessment? How so?

18 Unit 1 • Lesson 1.2

Lesson 1.3

Defogging with Questions

I'll never forget that one great time we had! Boy, that weather was something else! If it hadn't been for that one surprise, things sure would have turned out differently. But—who knew?

Hold on a minute. Who's doing all the work here? You're right—it's us, the readers! Do you find your head spinning with questions? Who is the narrator? Why was this time so unforgettable? What about the weather? What surprise? Foggy writing raises questions in a reader's mind. Good writers anticipate those questions and use clear details to answer them. Details have a wonderful way of defogging your writing.

An Example: Clear—or Foggy?

Read the following passage carefully. Then use the space right below to jot down any questions that occur to you. It's fine to confer with a partner as you work.

When I woke up, everything was strange and different.

A person wearing unusual clothes was looking down at me.

Ideas

Name _____ Date _____

My Questions:

1. _____
2. _____
3. _____
4. _____
5. _____

My General Response:

How did you feel, in general, about this piece?

☐ I had to do ALL the work! This writer barely told me anything.

☐ The writer really filled me in. I had trouble thinking of any questions.

A Second Example: *A Connecticut Yankee in King Arthur's Court*

In the classic story by Mark Twain, a man named Hank Morgan has been struck on the head in a fight. He awakens to find himself transported from nineteenth-century Connecticut to sixth-century England, the time of King Arthur and Camelot. As he comes to, Hank finds himself looking up at a rather unexpected face. As you read, notice the details Twain chooses to describe this unusual character.

When I came to again, I was sitting under an oak tree, on the grass, with a whole beautiful and broad country landscape all to myself—nearly. Not entirely; for there was a fellow on a horse, looking down at me—a fellow fresh out of a picture-book. He was in old-time iron armor from head to heel, with a helmet on his head the shape of a nail-keg with slits in it; and he had a shield, and a sword, and a prodigious spear; and his horse had armor on,

Ideas

too, and a steel horn projecting from his forehead, and gorgeous red and green silk trappings that hung down all around him like a bedquilt, nearly to the ground.

A Connecticut Yankee in King Arthur's Court
by Mark Twain

Making the Connection

Did you make the connection between the first and second examples? If so, you noticed that the first example is our foggy rewrite of Mark Twain's passage—we just took out all the details. Look at the questions you wrote following that foggy example. Circle the number of each one that was answered in Twain's actual passage.

Asking Readers' Questions

The following piece of writing could use some help with detail. Read it through one time to get a feel for it. Then close your eyes. Can you both see and feel what the author is writing about?

On one of the big holidays this year, the neighbors got a little out of control. They had more fireworks than last year. Some were really noisy and bright. I felt nervous all night. I was afraid something bad could happen.

What questions do you have as a reader? Write them here. (It's fine to work with a partner.)

My Questions:

1. _____
2. _____
3. _____
4. _____

Ideas

Using Questions to Revise

Now revise that short piece by answering your own questions. Add or replace any details you wish. Don't be afraid to invent! This is fiction, after all. Use your own experience to make the writing as vivid as if this were your own story.

Use your own paper or write on the original draft—your choice.

Share and Compare

Meet with a partner or writing circle to share your revisions by reading them aloud. What kinds of details did each writer come up with? Did everyone do a good job of answering a reader's possible questions?

Writing to Answer Questions

Now it's your turn to think of a topic, pose questions a reader might have, and write with detail to answer those questions. Begin by choosing a person to describe. It could be someone you know well—or a character from film or fiction. Here are a few suggestions:

- A friend from the present or long ago
- An unusual relative
- A coach or teacher you won't forget
- A pet or other animal important in your life
- Any fictional character you recall well

Once you've chosen your character, jot down three to five questions you think a reader would LOVE to have you answer about this character:

1. _____
2. _____
3. _____
4. _____
5. _____

Ideas

Write a clear, vivid description that answers these questions and any others that occur to you as you write. Use your own paper. Keep writing for 15 minutes or more.

Share and Compare

Before sharing, read your writing over aloud and softly to yourself. Insert any important details you may have forgotten. Then meet with a partner or writing circle to share your writing. As each writer shares, write any unanswered question you have on an index card, fold it, and hand it to the writer. Do NOT open any cards until everyone has shared.

A Writer's Questions

Some details are more critical than others. Is it important, for example, to know a character's height or eye color? Or do other things matter more? What kinds of details really help us to understand a character well? Can a story sometimes tell us as much about a character as a description? Why?

Putting It to the Test

Could listing questions to answer help you to make your writing more detailed in a testing situation? Would the quality of the writing depend on the quality of the questions? What could you do to make sure you asked very good questions?

Ideas

Lesson 1.4

Freedom from Filler

Ask anyone who's had that second (or third) piece of chocolate cake. Too much of a good thing can be as bad as too little. This applies to writing, as well. Shocking as it may sound, some writers overcrowd their writing with unneeded information just to make the writing longer. Their thinking is that longer is better. It *is*, right? Uh—wrong! In writing, more isn't better; only *better* is better. Details that are irrelevant or repeated are known as *filler* because they're only included to fill the page, not to entertain or enlighten anyone. Writers resort to filler when they're tired—or when they don't know the topic well enough to add anything new. Big mistake. Only add what's worth reading. Do your research, and keep your writing filler-free.

Sharing an Example: *Artemis Fowl*

Here's a short passage focusing on the title character from Eoin Colfer's futuristic fantasy tale, *Artemis Fowl*. Read it aloud carefully. Listen for information about Artemis and the world he lives in. What does this author want you to know? Read the passage a second time, pencil in hand. Underline details about Artemis and his world.

Ideas

After eighteen solid hours of sleep and a light continental breakfast, Artemis climbed to the study that he had inherited from his father. It was a traditional enough room—dark oak and floor-to-ceiling shelving—but Artemis had jammed it with the latest computer technology. A series of networked Apple Macs whirred from various corners of the room. One was running CNN's Web site through a DAT projector, throwing oversized current-affairs images against the back wall.

Butler was there already, firing up the hard drives.

"Shut them all down, except the Book. I need quiet for this."

The manservant started. The CNN site had been running for almost a year. Artemis was convinced that news of his father's rescue would come from there. Shutting it down meant that he was finally letting go.

"All of them?"

Artemis glanced at the back wall for a moment. "Yes," he said finally. "All of them."

Butler took the liberty of patting his employer gently on the shoulder, just once, before returning to work. Artemis cracked his knuckles. Time to do what he did best—plot dastardly acts.

Artemis Fowl
by Eoin Colfer

Ideas

Name _____ Date _____

What Did You Learn?

In this passage, the author is beginning to create a picture of Artemis and his world. In the box below, write at least five things you learned about Artemis or the world in which he lives. Include vital information and essential details only. If you think the author included any filler—distracting, unnecessary, or repetitive details—write that information outside the box.

Artemis and His World

_____
_____
_____
_____
_____
_____
_____
_____
_____
_____
_____
_____
_____

Ideas

Share and Compare

Meet with a partner to share what you learned about Artemis and his world. Did you include similar details? Did either of you find any *filler*? If so, be prepared to explain why you think it's filler.

Fill 'em Up

Sometimes writing badly on purpose makes you SO aware of a problem that you avoid it in your own writing. Ready to give it a try?

Meet with your writing circle to review the first paragraph of the passage from *Artemis Fowl* (printed below). We provided extra space between lines to give you room for adding filler. We also provided a couple examples using carets (∧) to show where we inserted filler. Work as a group to come up with as much filler as you can *without* changing the basic meaning. **Note:** You do not need to use our examples—change or eliminate them if you want. Have fun, but stay focused on the task: adding filler without changing basic meaning.

Waking slowly
∧ after eighteen solid hours of sleep and a light
            clothed in his Spiderman pajamas
continental breakfast, Artemis ∧ climbed to the study
            elderly, slightly balding
that he had inherited from his ∧ father. It was a

traditional enough room—dark oak and floor-to-ceiling

shelving—but Artemis had jammed it with the latest

computer technology. A series of networked AppleMacs

Unit 1 • Lesson 1.4   27

Ideas

whirred from various corners of the room. One was running

CNN's Web site through a DAT projector, throwing

oversized current-affairs images against the back wall.

Filler-Free Writing

Don't you feel better after getting that filler out of your system? Now the big question is how to keep it out—and away from your own writing.

To conclude this lesson, you'll write a descriptive paragraph, focusing on a place you know well or can imagine vividly. Here are some suggestions, but any place you can picture well will do:

- Your room
- A friend's or relative's home
- Anywhere with an amazing view
- The inside of your locker (or a friend's locker)
- A secret hiding place
- The bottom of the ocean
- Somewhere in outer space

Close your eyes for a moment and picture the place. Take it in: sights, sounds, smells, tastes, and feelings. Think what is most important—what your readers need to know to feel as if they're right there with you. Then—write. Write. *Write.* Don't stop until time is up. Use your own paper and skip every other line.

Ideas

Share and Compare

Read your draft over quickly and insert any last-minute details that occur to you. Then share your descriptive paragraphs with partners or in writing circles. Did you
- include critical details?
- answer readers' questions?
- avoid filler?

Work with your partner to do one last revision. Insert important details you missed. Cross out any filler. Read it aloud one last time to make sure you're satisfied with your revision.

A Writer's Questions

How do you know which details to cut and which ones to keep in a piece of writing? What sort of rules or personal system do you use? What happens when you cut too much?

Putting It to the Test

If you make your writing too short, won't you lose points in a writing assessment? If so, what should you do if you simply run out of things to say—repeat yourself (skillfully, of course) or just leave your piece short and hope for the best?

Conventions & Presentation

The WRITER...
edits everything thoroughly.

So the READER...
_____
_____
_____

The WRITER...
looks <u>and</u> listens for errors.

So the READER...
_____
_____
_____

The WRITER...
uses conventions to bring out meaning and voice.

So the READER...
_____
_____
_____

The WRITER...
is thoughtful about presentation.

So the READER...
_____
_____
_____

Conventions and Presentation
Editing Level 1: Conventions
Priority Access

It doesn't seem right to spend quality time developing and shaping an idea only to have the message lost through careless editing. The trouble with mistakes—in addition to the fact that they can be confusing—is that they're distracting. When readers stop to notice errors, they forget what you're trying to tell them. For your message to get through, you need to become a *careful editor*—with sharp eyes and ears, a grasp of conventional basics, and a willingness to work hard. Editing takes time but comes with big benefits—readers who understand your message and appreciate the trouble you took to make their job easier.

A Warm-Up

Before fine-tuning your editing, take a moment to assess your conventional skills. Reflect for a moment—then check (✓) your strengths and trouble spots on the chart below. When you finish, make some notes about things that give you particular trouble. (**Example:** If you check punctuation as a trouble spot, perhaps your particular challenge is semicolons.)

Personal Assessment	Strength	Trouble Spot
Spelling		
Punctuation		
Grammar		
Paragraphing		
Capitalization		

Specific things that give me trouble:

_____
_____
_____
_____
_____

Editing

Now let's warm up with some editing. Read the following example of writing carefully—and correct any errors you see. Some are tricky and hard to spot, so take your time (about 6 to 8 minutes). We recommend

- reading once for meaning and again to edit, pencil in hand.
- having a writing handbook or other resources nearby.
- using the correct marks from the copyeditor's poster.
- going back for one more look to be sure you didn't miss anything.

When you finish, record the number of errors you found.

It was the last weekend before school, and we hadn't been camping once. Usually my family take several camping trips, and I get too bring along couple of my friends. This summer, though my friends and I were all on the same baseball team, going to tornament after tornament. The problem was strange, as it may sound we kept winning—and all the camping time slipt away until we we're down to are last weekend. My Dad is the one who came up with the idea to get out the tent and set it up in the corner of hour yard. We jump at the ideas.

We got everthing set up inside the tent, but we didn't go in until it was dark. We could hear every sound from the neighborhood: cars, crickets, people talking, and even two cats chasing each other. It was hard to fail asleep untill we got used to the sounds. You could see constellations though the mesh netting at the top the top of the tent. Their were no cars going by; it seamed as though we were out in in the woods. It wasn't reel camping but it was close enough

I found _____ errors.

Share and Compare

Meet with a partner to share your work. First compare your personal strengths and trouble spots. Are any of your partner's trouble spots strengths for you—or vice versa? Are there ways you could coach each other?

Now compare your editing with a partner's. Did you find the same errors and the same number of errors? If you didn't, take one more look—this time, working as a team. Then coach your teacher as he or she models the editing of this passage.

From Simple . . . to Tricky

It's been said that "practice makes perfect," but the truth is closer to "perfect practice makes perfect." That's why you want to challenge yourself as an editor every chance you get. Read each of the following three examples carefully. Then edit—and rate the difficulty. You should notice the number of errors increasing as you go along.

Example 1

My sister is always dieting, and it makes me very worried! Their is really no reason to diet if you make sure you exercise and try to eat healthy foods Meghan doesn't want to exercise, and I think she is practically starving herself. I think she might have a kind of eating disorder, and should see a doctor. she doesn't want to listen to me, because I'm just her dumb little Brother

I found _____ errors.

Difficulty for me on a scale of **1** (piece of cake, dude) to **10** (I barely survived):

| 1 | 2 | 3 | 4 | 5 | 6 | 7 | 8 | 9 | 10 |

Example 2

I dont know what it is about my mom, but she's has alot of bad luck as driver. Let me give you few examples? To get to my drum teachers house, we half to go down a very busy street, the the kind that has a traffic light every block. Whenever were running a few minutes late, her bad luck kick in, and we became red light magnet I swear that some lights see us coming and switch from green to red without even a bit (even a hint of yellow in between.

I found _____ errors.

Difficulty for me on a scale of **1** (piece of cake, dude) to **10** (I barely survived):

1 2 3 4 5 6 7 8 9 10

Example 3

Moving from won city to another is hard but when the the new city is in different state a swell, moving can even be even harder. My Mom lost her job she has been looking for for work four about six month's. She told us not to worry, but my brother and I couldn't help but be nerveous. She told us, Its not your job to worry about my work. Just recently, she was offered a job, this was a real releif to the our family. The only catch was it was that this new job would means and so we had too move from Idaho, to Texas. Before we, made the big move, we flue to Texas too get a feel fore our new town and to serch for a house

I found _____ errors.

Difficulty for me on a scale of **1** (piece of cake, dude) to **10** (I barely survived):

1 2 3 4 5 6 7 8 9 10

Share and Compare

Compare your editing with a partner's. Did you find the same errors—and the same number of errors? When did the editing become difficult?

☐ Example 1—that was already a challenge for me!
☐ Example 2—I needed much more time on that one.
☐ Example 3—that's as much of a challenge as I can handle.
☐ It was never difficult. I'm ready for a harder challenge.

A Writer's Questions

Do most errors occur because the writer lacks skills in conventions or because the writer works too fast and doesn't pay close attention? Which one is closer to the truth for you? What does this tell you about improving your editing skills?

Level 2: Presentation
Keep It Under Cover

You know the old expression about judging a book by its cover, right? But do you think most of us do it anyway? It's not always literally a book we're judging. It could be a film, video game, magazine, newspaper, storefront window, Web site, or a hundred other things that use presentation to catch our eye or even get us to make a purchase. What's on "the cover"—what a reader or viewer sees first—matters because it makes a first impression that's sometimes hard to revise or erase. In this lesson, we'll focus on book covers, but think about how the features of a good book cover could improve other kinds of presentation as well.

Warm-Up

Together you and your teacher have gathered a small collection of books—picture books, novels, nonfiction works, and others. Create a display. Then with your partner, review the covers one by one. Use the writing space below to list the features you think should be considered in giving an award for the best book cover. We numbered through 6. Extend the list if more things occur to you.

1. _____
2. _____
3. _____
4. _____
5. _____
6. _____

Cover Design Guidelines

As a class, do three things:

1. Share the features you feel are important.

2. Narrow this total to a class list of six final cover design guidelines.

3. Brainstorm the name of an award you might, as a class, give to the book cover you felt did the best job of following those guidelines.

Share and Compare

When you finish, use the guidelines to review the book covers one more time, working with your partner. Together, nominate one as Most Outstanding. List all the nominations for your class and take a vote. Present your award by

- displaying the book,
- creating an actual award on paper, OR
- writing to the author or publisher to express your approval.

Design Editor for a Day

In this portion of the lesson, you get to be the design editor—just for today. Your task is to choose or design a cover for a new bestseller titled *Moose Creek Memories: Growing Up in the Wilderness*. The author, Pete Moss, has had several books on *The New York Times* bestseller list. (Sorry—both book and author are hypothetical.)

You will work with your writing circle, and we will offer you some preliminary suggestions, but you are not bound by any of them. Feel free to go with your own original ideas. Before looking at our proposed designs, brainstorm your group's original ideas and make some notes here:

Book cover ideas . . .

_____
_____
_____
_____
_____
_____
_____
_____

Three Possibilities . . .

Following are three book cover proposals from designers who'd love to have their work chosen for a new book by a best-selling author. Each designer hopes to have THE cover that will motivate book buyers to pick up his or her book or make that online click to purchase.

Does one catch your eye more than the others? Or is there one you could combine with your own design ideas? Which one will speak to readers?

Design 1

Design 2

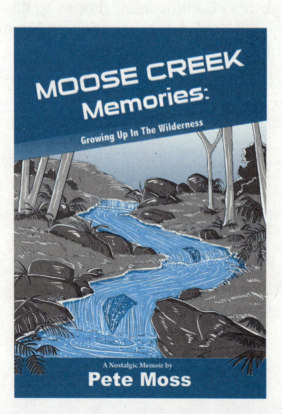

40 Unit 1 • Conventions & Presentation

Design 3

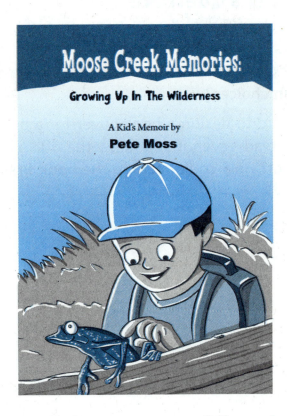

After looking at all three covers carefully, make your choice—and talk about other design issues, such as font, color, placement of art or title, etc. Use this space to make some notes.

We chose

☐ Cover 1
☐ Cover 2
☐ Cover 3
☐ None of them—we prefer our own design ideas.

What other design issues or concerns are important? List them here:

1. _____
2. _____
3. _____
4. _____
5. _____

A Writer's Questions

So much of today's book shopping and buying occurs online. When a reader can't actually touch or pick up the book, how does this affect the cover design choices?

Is it hard to design a book cover for a book you've never read—even if you know the title and basic content? How could reading the book first help?

Presentation Matters

Imagine that you have been given the job of creating the cover for a new book or other work. We urge you to think of your own idea, but we'll also offer you two titles to choose from, if you prefer. (If you are working on a story, report, or other piece right now, you might choose to make a cover for that.) Your cover will need to include these design elements

- Title
- Author's name (you or an author you invent)
- Art
- Choice of color
- Choice of font
- Suggestions for placement and spacing

NOTE: If you're a great artist, terrific! Here's a chance to use that talent. But if you're not, don't worry. What matters most is for you to communicate with the artist who WOULD be on your editing team if this book or article were actually going to press. Do a rough sketch or use words to describe what you would like to see. Remember to be specific about styles, size, and color. You may also borrow non-copyrighted art online. Cite your sources.

Here are two title suggestions if you do not have a title in mind:

- *Vampire Stories Are So Last Year* by Wolf Mann
- *The ABC's of Web Design—How Anyone Anywhere Can Design an Amazing Web Page* by Itso Simple

HINT: Make a rough sketch on scratch paper first. Then create your actual cover on any paper of your choice. If you have computer access, by all means design your cover digitally. Remember to give proper credit for any art you borrow, and be careful not to import art or designs that are copyrighted. Speaking of which—DO remember to copyright (©) your own work!

Ideas

Sample Paper 3

Score for Ideas _____

Seahorses

The seahorse is, of course, not a horse at all—and it isn't mythical, either. It does, however, have connections to both horses and myths. The head of this delicate and unique creature has a striking resemblance to the head of a horse. Its scientific classification, the genus Hippocampus, is named after a figure from Greek mythology who was half horse and half fish. And there's much more about this tiny sea creature that is both strange and remarkable.

Unlike most fish, the seahorse has no scales. Its skin is stretched tight as a drum over bony plates beneath—and can change color to blend with almost any surroundings, providing this little creature's only real defense against predators. The seahorse uses its long, curly **prehensile** tail to grasp onto ocean plants when it wants to hide from enemies or when it needs to stay in one place to eat. Grabbing onto a plant and anchoring itself frees the seahorse to use its long snout in capturing any food that happens to drift by. Though tiny, seahorses are voracious carnivores, feasting on crustaceans and plankton—all day if possible—to satisfy their huge appetites.

Seahorses are constantly famished because they burn so much energy just getting around. Their feather weight and upright body position makes swimming extraordinarily difficult. To push through the water, seahorses use powerful dorsal fins that beat almost as fast as a hummingbird's wings. During a storm, when waves are huge and ocean currents are strong, seahorses can wear themselves down to the point of total exhaustion or death.

Everything about the little seahorse is unusual. They are **monogamous,** having only one partner for life. Perhaps most startling of all, it is the male that gives birth! According to *National*

Geographic, seahorses are the only animal species on the planet for which this is true. The female deposits eggs into the male's **brood pouch** where they are fertilized, and after 10 to 30 days, the male gives birth to as many as 1,000 fully formed seahorse babies. This may seem like a lot (and it is), but very few of these babies reach adulthood. Most are eaten by fish and turtles—or captured by humans.

Seahorses need our help and protection. Pollution, shrinking habitat, and over-harvesting (mostly for exotic aquariums) are leading some species of seahorse towards extinction. Seahorses rarely survive in captivity, but sadly, this only increases the demand. China, Taiwan, and some parts of Europe have set up sanctuaries for the seahorse or made laws outlawing its capture. By taking steps to minimize ocean pollution and by opposing the display of this frail sea creature in aquariums (where they are doomed), we can extend their existence.

Glossary

- **brood pouch**—a pouch where young are held and protected during early stages of development.
- **monogamous**—the habit of having only one mate at a time.
- **prehensile**—capable of grasping.

Sources

George, Twig. *Seahorses.* Brookfield, CT: Milbrook Press, 2003.

Wallis, Catherine. *Seahorses.* Guilford, CT: Bunker Hill Publishing, 2005.

NOVA: Kingdom of the Seahorse. DVD production. 1974. WGBH Boston. Released June 6, 2006.

http://animals.nationalgeographic.com/animals/fish/sea-horse.html

Ideas

Name _____ Date _____

Sample Paper 4
Score for Ideas _____

Energy—A Problem

The world's population cannot keep on using energy sources like gas made from fossil fuels, oil, and coal. If we do, we will not be able to drink our water or breathe our air. These things cause all kinds of pollution.

Traffic is getting worse all the time. But mostly, people worry about the price of crude oil and the price of gas at the pump. It might actually be a good thing if gas prices were higher. People might actually decide to drive less, which would help reduce at least one kind of pollution. The average American, in particular, drives far too many miles in a given year.

Energy conservation goes beyond the gas pump. Turning off lights and appliances would help, as would turning off computers when they are not in use.

Solar power could help because it does not create the same pollution as oil or gas, and it is free (the sun's rays are free though the solar equipment needed to capture this energy is not free). We could also use wind power. This is a good solution in many parts of the world.

Some day, we might have cars that efficiently use other kinds of fuel, such as hydrogen.

It is important to work together to solve the Earth's energy problems. If we don't, these problems will only grow worse. No one wants that.

Sources

Smil, Vaclav. *Energy: A Beginner's Guide.* Oxford, UK: Oneworld Publications, published in 2006.

Christine Peterson. *Alternative Energy.* Danbury, CT: Children's Press, 2004.

http://earth911.com/reduce/energy-costs-and-conservation-facts/

Ideas

Name _____ Date _____

Revising Checklist for Ideas

- ☐ I chose a writing topic I like and I'm excited about it. OR . . .
- ☐ I want to change my topic to _____.
- ☐ I have all the information I need to make writing easy. OR . . .
- ☐ I could get more information from _____.
- ☐ My main idea is focused and manageable. OR . . .
- ☐ I'm going to shrink it down to this: _____.
- ☐ Details expand my discussion or story for readers. OR . . .
- ☐ I need to include details that answer these readers' questions:

  _____?

  _____?

  _____?

- ☐ I crossed out any filler (unneeded information that interrupts the main message). OR . . .
- ☐ I didn't have any!
- ☐ My title hints at the main idea without telling *too* much.
- ☐ _____ rated my writing for Ideas:

1 2 3 4 5 6

> **Note** Don't use this checklist just to compliment yourself—even if your writing is terrific! Use it to plan revisions or additions. Try to see your writing the way a reader who doesn't know you would see it.

Ideas

Name _____ Date _____

Revising Checklist for Conventions and Presentation

☐ I waited at least 3 days to edit my draft so I could see it "fresh."

☐ I read my writing twice to check for errors—once silently, once aloud.

☐ I looked carefully at these things: ___ capitalization ___ spelling ___ grammar ___ punctuation ___ paragraphing

☐ There are NO distracting errors (even tiny ones) to slow a reader down or get in the way of the message.

☐ I used punctuation to bring out meaning and voice.

☐ I used *italics* to show readers which words to *emphasize* aloud.

☐ I used **boldface** to make important terms stand out.

☐ IF I used dialogue, I started a new paragraph for each new speaker, and

☐ I used quotation marks to mark each speaker's words.

☐ I designed my presentation to catch a reader's eye.

☐ My presentation makes the "informational trail" easy to follow.

☐ I made important information (facts, names, dates) easy to find.

☐ My presentation makes my message easy to understand, remember.

☐ _____ rated my writing for Conventions and Presentation:

| 1 | 2 | 3 | 4 | 5 | 6 |

Note Good conventions and presentation are a matter of courtesy. After all, *someone* edits every piece of writing that is read. The question is, will that someone be you—the writer? Or will you leave the editing task to your reader?

48   Unit 1 • Checklist

UNIT 2
Organization

Organization is one of those things we don't always appreciate until it's missing. Imagine a world, for instance, in which maps didn't match real-world geography, clocks and calendars operated randomly, keyboards mysteriously shifted daily so you could never predict where the letters would be, and pages of books were shuffled like cards, making you scramble to read things in order. Luckily, a great number of things in our lives are meticulously organized. We want your writing to be one of them.

In this unit, you will expand your understanding of Organization by

- choosing an organizational design to suit your purpose.
- beginning good organization with a solid main idea.
- using three anchors—lead, conclusion, transitions—to make organization strong.
- organizing details to create a strong informational paragraph.

Organization

Sample Paper 5
Score for Organization _____

The Monster

My best friend K.C. said she wanted to go to the amusement park. This was on a Saturday, and there was not that much to do anyhow. I didn't have a babysitting job that day, so I decided to go with her.

We just walked around for a while and ate cotton candy and elephant ears and stuff like that. It was so fun! I love all the smells of junk food and all the noise at the amusement park. I kept thinking we might meet some friends, but we didn't. K.C. wanted to go on the roller coaster. I didn't want to because I am terrified of heights, and wild rides make me sick. I can't even climb ladders or trees. Coming down is even worse than going up.

We wandered some more, hoping to bump into friends, and K.C. called some people on her cell phone. She also called her mom. Then it was almost time for us to leave, so I told her I would ride the coaster. We were strapped in, and then we were flying through the air. The only other time I was that scared was when I was eight and my mom took me skiing. I got going down a really steep hill and couldn't stop. It totally freaked me out.

Later, K.C. said I was screaming the whole way. Well, I don't think I was, but I don't remember any of it. K.C. wanted us to ride again, but I told her to forget it. So she rode by herself. I think someday I might go on that ride again, but maybe not. K.C.'s mom drove us home after that. I thought I was going to get sick in the car. I didn't, though, thank goodness. By the way, in case you haven't guessed it, the ride was called The Monster.

Organization

Sample Paper 6
Score for Organization ____

Camping

What is the most overrated activity, outdoor or indoor, in all of America? If you said, "camping," or even thought the word, you win the big prize. Don't get me wrong. I like being outdoors, and I even like hiking for short distances (with plenty of water). I just don't want to live outdoors, not even for a weekend. If you are one of those people who really enjoy camping, that's OK. I say, go for it! Just don't invite me.

The problem with camping starts with the packing. My dad, of course, loves to camp (obviously, it's not genetic). He always says, "Bring only what you need." Is he kidding? As if it's that simple. Everything I need is right here at home! What he really means is, "Bring only what fits into one small backpack," and it's a pack that *I* will be carrying over miles and miles of primitive, rough trails. If I had a pack mule to haul my television, computer, and my bed with all my pillows and blankets (or at least a sleeping bag long enough to let me unbend my knees), I would be much happier. As it is, I can only squeeze in two T-shirts, some socks, an old toothbrush, and a deck of cards—at least these things don't weigh much.

The trouble continues once we hit the trail. This is the part where we hike to our campsite. Other people *drive* to campsites where they pitch tents in shady spots near a beautiful river. (Even in TV advertisements you see people drive their cars to campsites.) As we approach the trailhead, I see people already parked and enjoying life, and it always makes me jealous. They are sitting in chairs cooking their hot dogs, laughing as though camping is the greatest thing ever. That's not for us. They're wimps. We're pioneers. Everyone knows the best campsites are in the remote wilderness. That's what we set out to find—the more remote the better. If your compass is still working, it's not remote enough yet. You have to arrive at your campsite scratched and bleeding from hiking through prickly underbrush, so dehydrated and exhausted you can barely pitch your tent.

But later . . . when we are finally around our own campfire roasting marshmallows (actually, setting them on fire) and listening to the wind in the trees (wondering if it's a cougar), I *almost*, repeat *almost* like camping. I know that I won't see a real bathroom for days, every muscle in my body will ache, my hair and teeth will get grittier by the hour, and I'll have to endure being tortured by 10,000 mosquitoes. But seeing the smile on my dad's face makes me feel good. (Don't tell him I said this. I'll just deny it.)

Organization

The WRITER...
opens with an inviting lead.

So the READER...
_____
_____
_____

The WRITER...
organizes information to showcase the message.

So the READER...
_____
_____
_____

The WRITER...
uses helpful transitions.

So the READER...
_____
_____
_____

The WRITER...
closes with a satisfying conclusion.

So the READER...
_____
_____
_____

Unit 2

Organization

Name _____ Date _____

Lesson 2.1

Finding the Right Design

In every police drama, detectives hot on the trail of a criminal scrutinize every clue, seeking a pattern to help them solve the case. The pattern, or organizational design, of a document can also help readers in a similar way, offering clues about how the writer has ordered or grouped ideas. A writer selects an organizational design to fit his or her topic and purpose. A description of a Gila monster calls for one organizational design, a recipe for lasagna quite another. Longer pieces of writing may call for a creative blend of designs that allow a writer to shift purposes, mixing narrative, informational, and descriptive writing, for example. In this lesson, you'll become familiar with five organizational designs and choose the one to fit a piece of your own writing.

Five Popular Designs

Following are short descriptions of five widely used organizational designs. There are, of course, many more. As you read each description, use a check mark (√) in the box to show if you've used the design in your own writing or noticed it as a reader.

Design 1

Chronological Order: This design arranges information or events based on a progression of time, sometimes flashing back or leaping ahead.

Examples: *stories and novels, history, news stories, biographies*

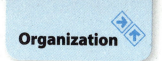
Organization

Name _____ Date _____

☐ I have used this design in my own writing.
☐ I have noticed this design as a reader.

Design 2

Spatial Order: This design arranges details in terms of place. It is useful for creating images or "moving pictures" in a reader's mind, helping the reader to see people, animals, objects, landscapes, and places as tiny as a single cell or vast as the universe.

Examples: *descriptions of people, objects, living things, places or landscapes, settings for a story*

☐ I have used this design in my own writing.
☐ I have noticed this design as a reader.

Design 3

Climactic Order: This design lets the writer lead up to (or sometimes wind down from) the most important point or event that writer has to share.

Examples: *news stories, literary analyses, major historic events*

☐ I have used this design in my own writing.
☐ I have noticed this design as a reader.

Design 4

Cause and Effect: Writers use this design to show how two events or behaviors are connected—specifically, how one thing causes or leads to another.

Examples: *persuasive essays of all kinds, predictions, critical analyses of current events*

☐ I have used this design in my own writing.
☐ I have noticed this design as a reader.

Organization

Name _____ Date _____

Design 5

Comparison-Contrast: Writers use this design to show how two people, objects, living things, places, events, concepts, and so forth are alike or different.

Examples: *persuasive essays, advertisements, literary analyses, historical analyses, political promotions, descriptive analyses*

☐ I have used this design in my own writing.

☐ I have noticed this design as a reader.

Reflection

Think about the kinds of writing you have done (or encountered) either outside of school or in your science, social studies, math, or health classes. Based on your experience, can you think of any organizational designs that are not included on our list?

List them here:

1. _____
2. _____
3. _____
4. _____

As you make notes, it is fine to

- check with a partner or other people in your writing circle.
- look through any books or other publications in your classroom.

Share and Compare

Share any additional designs you thought of and make a class list.

Organization

Name _____ Date _____

Name that Design

Read each of the following examples carefully, and see if you can identify the organizational design the writer used. Write the name of the design in the space provided and be prepared to explain your choice.

HINT: It helps to use an active reading strategy, circling important words or phrases that give clues about the design.

Example 1

My dad has become a runner. When he started, he could barely run to the end of the driveway. But he persisted, walking and running four miles daily. Each time, he would run as far as he could and walk when he was out of breath. He never gave up. Now he runs the full four miles, and he says he's setting a new goal: six miles. He hopes to run ten miles by the end of summer. He lifts weights, too. As a result, he's dropped thirty pounds and has had to buy all new clothes. Even his shoes got big! Who knew that could happen?

Organizational Design: _____

Example 2

Last night's thunderstorm put a local swimming pool out of business—at least temporarily. Guests and staff at the Lazy River Inn woke up to discover that the electrical storm had knocked out the pump on the resort's largest pool. According to a resort spokesperson, lightning struck the pump at about 1 a.m. Luckily, it did no other

Unit 2 • Lesson 2.1   57

Organization

hour. This morning, however, eager guests who had not heard the news were lined up outside the pool well before 10 a.m., the usual opening time. Temperatures were already close to 90 degrees, and a wave of disappointment swept over the swimsuit-clad crowd as the bad news was announced over the PA system. The Lazy River Inn hopes to have the pool up and running within two days.

Organizational Design: _____

The Right Design for the Job

Following are three writing tasks. Talk with a partner or members of your writing circle to choose the best design (or designs—it could take more than one) for each task. Write your choice in the space provided. Don't choose too quickly. Try to picture what the final piece of writing might look like.

Task 1

An essay about the devastation following Hurricane Katrina

Organizational Design: _____

Task 2

An editorial supporting a change in the school's mascot and team names

Organizational Design: _____

Task 3

An essay comparing a novel with the film adaptation of that novel

Organizational Design: _____

Organization

Name _____ Date _____

The Right Design for You

Choose a topic you could write about right now, without doing any research. Write about anything that's on your mind, or use our list to help you think of an idea. Follow these four steps:

1. Choose a topic (your own or one from our list below).

2. Choose a design that's a good fit. (It does not have to be one from this lesson.)

3. Do some prewriting, making notes that will help you follow your design.

4. Write for 15–20 minutes, making your ideas as easy to follow as possible.

My Topic

★ My topic: _____

- Being a teenager now versus in the past
- Description of any outdoor scene
- Your first day in kindergarten (or first day anywhere)
- Surprising things people learn playing video games
- Results of yielding to peer pressure

Organization

Name _____ Date _____

Share and Compare

Meet with your writing circle to share and discuss your writing. As you listen, think about each writer's design, but above all, ask this question: Is this writing easy to follow?

A Writer's Questions

Many writers feel most comfortable with a chronological order design. Why do you think this is? What problems could a writer encounter if he or she were comfortable with only one design?

Putting It to the Test

Do you think prompt writers have a design in mind when they write their prompts? Would it be helpful to figure out what that design is and follow it?

Organization

Name .... Date ....

Lesson 2.2

Putting Ideas First

All writing starts with the trait of Ideas. Without a focused message, it won't matter how much effort you put into the other traits—Organization, Voice, Word Choice, Sentence Fluency, or Conventions and Presentation. If you think this sounds like an exaggeration, imagine yourself trying to repair a crumbling bridge. Where would you begin? With paint? A new railing? A flashy canopy? Or the basic structure? Writers too often leap right over the step of defining the message, thinking they can fix everything with a snappy introduction, a punchy conclusion, flowery language, or a little reordering of information. They're painting the bridge even as it collapses under their feet. Put the trait of Ideas first, and you'll have a solid structure that makes every other trait, including Organization, fall right into place.

Finding the Source of the Trouble

Following are two short examples of student writing. Each needs revision. The question is: What's the problem? Is it the ideas, the organization, or a little of each?

Read each example carefully, asking the following readers' questions as you go.

- Is there a big idea?
- If so, what is it? If not, what should it be?
- Does the writer need more information?
- Is there a logical, ordered flow to the ideas?
- Is there filler that should be cut?
- Do the lead and conclusion work?

Organization

Name _____ Date _____

Use the space between lines to make notes. Then with a partner or writing circle, identify the big idea (if there is one) and jot down the three revision tips you would give to each writer.

Example 1

  Swimming can make you extremely fit. It's a good idea to begin young—four or five years old, or even younger. Young children have far less fear of the water. Very young children can be hard to teach because they have difficulty following directions. Everyone should know how to swim. Each year, people get into trouble because they panic and do not wear life vests. Boating accidents are more common on rivers than anywhere. Most people who swim for exercise work out in a pool. It takes about thirty minutes of practice a day to maintain a good fitness level. Lifting weights is helpful, too.

Big Idea: _____

Our revision tips for the writer:

  1. _____
  2. _____
  3. _____

Organization

Name _____ Date _____

Example 2

Some people never travel out of the state where they were born. Travel can be expensive. It can be dangerous, as well. Some schools offer travel on a limited basis, through field trips or summer trips. Some are sponsored by local businesses, or students earn part of the money through school activities. Imagine if students did not begin college right away, but instead they had an option to travel for two years. Where would they go? That could take some thought and planning. If you are going to travel, it is good to be organized. Travel is highly educational.

Big Idea: _____

Our revision tips for the writer:

1. _____
2. _____
3. _____

Share and Compare

Share your responses with the other writers. Did you identify the same big ideas, or was it difficult to pick out any big idea? Did you identify the same problems? Discuss any revision strategies you have for these writers.

Organization

Take One On

Now you'll choose one example to revise. Work with a partner and complete the following steps.

1. Read the example aloud a second time.
2. Look at any notes you wrote.
3. Identify the big idea, or decide what it should be.
4. Look over your three revision tips for the writer.
5. Based on your notes and tips, make a revision plan.
6. Then write your revision on scratch paper.

Share and Compare

Meet with a partner or your writing circle to share your revised examples. Even though you worked together in planning your revision, do your drafts look and sound different? Who has a strong lead? Conclusion? Message? Design?

A Writer's Question

How or why does the trait of Ideas pave the way for the trait of Organization?

Putting It to the Test

Let's say that a writer has perfect conventions, a wonderful vocabulary, brilliant sentence structure, and a comical voice in his or her writing. Will those things make up for not having a strong main idea?

Organization

Name _____ Date _____

Lesson 2.3

Holding It Together

Organization takes much more than putting details into a neat, orderly list. It is a way of holding little ideas together so they form something bigger, such as an essay, story, poem, argument, and so on. Think of it this way: Suppose we use bricks to represent individual details. We can organize those bricks by stacking them to make a wall, but that wall won't be very strong without some mortar to hold the bricks in place. Three organizational features make a writer's wall of ideas particularly strong—the lead, the conclusion, and transitions that connect ideas. In this lesson, we'll look at them one by one, and then see how they work together.

Sharing Favorite Leads

A good lead is especially important because it kicks off any piece of writing. It's the first thing the reader sees or hears. Meet with your writing circle to share some favorite leads. Then, together, write down three qualities of a good lead.

1. _____
2. _____
3. _____

Organization

Choosing the Lead that Works

How many leads do you usually write for one piece? Would it surprise you to hear that professional writers often write a dozen or more? Following is one lead from a professional writer, Ji-Li Jiang, author of *Red Scarf Girl*. We grouped it with two others that we wrote. See if you can tell which one of the three is Jiang's lead. Put a check by it.

☐ I remembered coming home from kindergarten and showing Grandma the songs and dances we had learned. Grandma sat before us with her knitting, nodding her head in time to the music.

☐ I have wonderful memories of my grandma. She was a small, very friendly person, who made friends easily. She loved music and knitting.

☐ My grandma was a really special person. She always had time for us, even when she was busy with her household chores, such as knitting.

Exploring Transitions

Transitions are words and phrases that link ideas, thoughts, and, sometimes, whole paragraphs together. You no doubt use them all the time, but you may or may not refer to them as *transitions*. Some people call them linking words or word bridges.

Meet with your writing partner or writing circle. Working together, create a list of transitional words and phrases organized by purpose. We have given you a few words or phrases for each category. See if you can add one or two more.

Organization

Transition Words

- Words to show **time:** *while, meanwhile, next week, afterward, then, suddenly*

    Other words we thought of: _____

- Words to set up a **comparison:** *likewise, also*

    Other words we thought of: _____

- Words to set up a **contrast:** *although, but, nevertheless*

    Other words we thought of: _____

- Words to create **emphasis:** *especially, for this reason*

    Other words we thought of: _____

- Words to **wrap things up:** *in conclusion, finally, anyway, in the end*

    Other words we thought of: _____

- Words to **add information** or **set up an example:** *for instance, also, and*

    Other words we thought of: _____

Sharing an Example: *Red Scarf Girl*

Let's put these first two organizational pieces together with an example passage from author Ji-Li Jiang's memoir, *Red Scarf Girl*. In this book, Jiang tells about growing up in China during Mao Zedong's Cultural Revolution. Her family, shunned by neighbors and former friends, lived in constant fear of harassment or arrest. In your writing circles, read the following passage describing Ji-Li's grandmother, who came to live with them to escape trouble. Follow these steps:

1. Notice the lead. Did you choose it from the previous list of three?

2. Using your transition list as a reference, highlight any transition words or phrases you find in the passage.

Organization

Name _____ Date _____

I remembered coming home from kindergarten and showing Grandma the songs and dances we had learned. Grandma sat before us with her knitting, nodding her head in time to the music. Sometimes we insisted that she sing with us, and she would join in with an unsteady pitch and heavy Tianjin accent, wagging her head and moving her arms just as we did.

When we tired of singing, we would pester Grandma to show us her feet. When she was young it was the custom to tightly bind girls' feet in bandages to make them as small as possible—sometimes as small as three inches long. This was considered the height of a woman's beauty. Grandma's feet were half bound, and when she was only seven she fought to have them released. As a result her feet were smaller than natural feet but larger than bound ones. We loved to touch them and play with them. If she refused to let us, we would tickle her until she panted with laughter.

Red Scarf Girl
by Ji-Li Jiang

Share and Reflect

Share the transitions you found with the rest of the class. What did you notice about this author's use of transitions?

☐ She used so many it was actually confusing and overwhelming to me.

☐ She used so few I had a hard time connecting ideas.

☐ Her use of transitions was very balanced—just enough to create a smooth flow of ideas.

The Finishing Touch

This passage from Ji-Li Jiang actually has a conclusion, but we left it out to give you a little challenge. With your partner or writing circle, complete the following steps.

Organization

Name _____ Date _____

1. Read the passage aloud, one more time.

2. Talk about ways the writer might wrap up this passage (not the whole book—just this part).

3. Draft your own conclusion, about two to four sentences.

Putting All the Pieces Together

It's time for you to put the pieces together in your own writing. There's a lot to think about with good organization, but you can do it. Begin by listing the important organizational features you remember from this and past lessons.

1. _____
2. _____
3. _____
4. _____
5. _____

My Topic

Now choose a topic for a paragraph or two—something that's on your mind right now or an idea from our list.

★ My own idea: _____

- An early memory
- Country music vs. Rock or Hip-hop
- Consequences of rude behavior
- What you'd see from my rooftop

Unit 2 • Lesson 2.3   69

Organization

Write for 15 minutes or more. In your mind, imagine the reader saying, "What's your big idea?"

Share and Compare

Meet with a partner or your writing circle and take turns sharing your drafts aloud. Listen for the following strong organizational features.

- A clear central message that makes it easier to organize
- A design that's easy to follow
- A strong lead
- Good transitions to connect ideas
- An effective conclusion to wrap things up

Pick one example to share with the whole class.

A Writer's Question

Organization seems to have so many parts. How can a writer possibly remember to include them all?

Putting It to the Test

Organizationally, what's the biggest mistake a writer can make in an on-demand writing situation?

Organization

Name .... Date ....

Lesson 2.4

The Whole Package

As we've seen, organization has many pieces and parts. In this lesson, you'll get to put everything together in a piece on earthquakes. What's that you're saying? You'd prefer a different topic? Not to worry. You can choose a different topic altogether, but freedom comes with a price—you'll need to do your own research. We've already dug up numerous details about earthquakes, and we're willing to share one or two of them. Just kidding. We'll share them all. So, the choice is yours. Your job is to create a draft that fascinates readers; those who already know about them and those who are reading about earthquakes for the first time.

Shopping for Details

Imagine yourself at an information shopping mall, pushing a shopping cart labeled MFT (My Focused Topic). You enter the store called Earth and find the section on Earthquakes. All the following facts and details are available for purchase. Which facts and details will you put into your cart? Think about your readers. If they were with you, they'd say, "Pick the *interesting* stuff—things we don't already know!"

Put a check in the box beside the items you want.

WARNING: Are you one of those shoppers who can't resist *anything*? You could be in trouble. If you pick up more than half the items on this shelf, you might make your writing task too big to handle!

Unit 2 • Lesson 2.4   71

Organization

Name _____ Date _____

Twenty Tidbits on Earthquakes

- [ ] 1. Earth is formed of several layers.
- [ ] 2. An earthquake is the often severe vibration of the surface of Earth that comes after some kind of energy release in Earth's crust—volcanic eruptions, human-made explosions, or the plates that make up the outer layers sliding over or under one another.
- [ ] 3. Faults are breaks in Earth's crust.
- [ ] 4. Earthquakes often recur along faults.
- [ ] 5. Earthquakes can be very destructive.
- [ ] 6. Landslides caused by earthquakes are often more destructive than the earthquakes themselves.
- [ ] 7. Earthquakes happen in many parts of the world.
- [ ] 8. If an earthquake originates below the ocean's surface, it can create huge waves called tsunamis.
- [ ] 9. Seismographs are instruments that detect, record, and measure vibrations caused by earthquakes.
- [ ] 10. Information from seismographs is often broadcast on your local weather channel.
- [ ] 11. The Richter scale measures the magnitude of earthquakes. A magnitude of 2.0 is the smallest quake usually detected by people; magnitudes of 6.0 or more are considered major.
- [ ] 12. The Richter scale is named after Dr. Charles F. Richter of the California Institute of Technology.
- [ ] 13. The U.S. Geological Survey does research on the likelihood of future earthquakes.
- [ ] 14. The largest earthquake of the twentieth century measured 9.50 and occurred off the coast of Chile in 1960.
- [ ] 15. A tsunami from the Chilean earthquake hit Hawaii, killing 61 people.

72 Unit 2 • Lesson 2.4

Organization

Name _____ Date _____

☐ **16.** If you are inside a building during an earthquake, it is usually safer to stay inside and take cover underneath something sturdy, like a doorframe.

☐ **17.** During an earthquake, many people are injured by falling debris from buildings and from downed electrical lines.

☐ **18.** California, Nevada, Idaho, Montana, Washington, Hawaii, Arkansas, Missouri, and Alaska have each had one or more earthquakes of magnitude 7.0 or higher in the last 200 years.

☐ **19.** Earthquakes in 1811 and 1812, near the border of Arkansas and Missouri, changed the course of the Mississippi River and even forced the river to flow backward for several hours.

☐ **20.** A major 7.0 earthquake struck near the Caribbean city of Port-au-Prince, Haiti in January 2010. Within two weeks of the quake, over fifty aftershocks measuring over 4.0 had been recorded. An estimated three million people were affected.

Reflection

Take a look at the items in your shopping cart. Can you group them together to make one big idea about earthquakes? Which of the following is true for you?

☐ I overloaded. I need to take some items out.

☐ I shopped too fast. I don't have enough information.

☐ I have just enough information, and my details go together well.

Organization

Name _____ Date _____

What's the BIG Idea?

You should be able to make one big idea or point about earthquakes. (If some items don't fit, toss them out.) State your main idea or message here:

_____

_____

HINT: The single word *earthquake* doesn't work as a main message. (You know why!)

Choosing a Design

With your narrowed big idea firmly in mind, choose an organizational design so you can begin to get a vision of how your writing will look. (It may very well change as you write. This often happens.)

- [ ] Chronological Order (time)
- [ ] Spatial Order (description)
- [ ] Climactic (leading up to a major point or points)
- [ ] Cause and Effect (how one thing causes another)
- [ ] Comparison-Contrast (how things are alike or different)
- [ ] Other _____

With your design in mind, number your details in the order you plan to write about them. You're not locked into this order—it too can change as you write.

Start with a Super Detail

Dip into your research once more. Find the one detail that you know will grab your reader's attention. Use that super detail to draft one possible lead on scratch paper. (You may wind up writing two or three leads and choosing the best one.)

74   Unit 2 • Lesson 2.4

Keep the Energy Flowing

Building on your lead, dive in. Write for 15 minutes or more, sharing the best of your details about earthquakes.

HINT: Remember to save one super detail for your conclusion. Use your own paper.

Share and Compare

Meet with your partner or writing circle to share your writing aloud. Listen for the following.
- Strong lead
- Strong conclusion
- Details that work together to create a main message
- Good design—easy to follow

Choose one paper to share with the whole class.

A Writer's Questions

We've seen that the trait of Ideas provides a good foundation for the trait of Organization. But can it also work the other way? How does the trait of Organization support the trait of Ideas?

Putting It to the Test

What features of good organization are particularly important in on-demand writing?

Conventions and Presentation
Editing Level 1: Conventions

Readers Need a Comma Break

Do you like pizza, tacos, burgers, fried rice, or all four? If that question makes you hungry, we apologize.

The answer doesn't matter, by the way. We just wanted to show off two uses for the comma, a very handy punctuation tool for organizing words within a sentence. Look at those first two sentences again. Can you describe one use for the comma—or both?

You've been using commas for a long time now, both when you write and when you speak. In fact, you've probably internalized many comma rules without even knowing it. The word *internalized* means you use them without even thinking about it, even if you couldn't recite the rule.

With this lesson, we're going to focus on the following two uses of the comma.

- Commas in a series
- Commas after an introductory clause (or introduction)

We used them both ways in the first two sentences of this introduction! We'll also focus on one specific way never to use a comma. (More on that later.)

A Warm-Up

Before we get into editing, let's look at the wide range of ways writers use commas. Choose a book to browse through. It can be anything except poetry. (Poets are sometimes free spirits when it comes to commas.) Find one sentence with a comma that you could share with your writing circle.

Write down the book title and author. Then write the sentence with the comma or commas on the lines below. Circle one of the commas. Make a guess about why the writer used it. What purpose does it serve? What does it tell the reader about reading the sentence?

Book Title: _____

Model Sentence:

_____

_____

My thoughts about what this comma shows:

_____

_____

Quick Question

If you wanted to find out more about commas and how to use them in your writing, where could you look for help—specifically? List three sources:

1. _____

2. _____

3. _____

Share and Compare

Share your sentence and your thoughts about the rule with your writing circle. If you're not sure about a rule, look at one of the sources you identified for help.

Three Important Comma Rules

Now let's focus on just three rules. These are simple, but very important.

Unit 2 • Conventions & Presentation

Rule 1

- Use a comma after an introductory clause or phrase.

    Examples

    If you want your dog to learn to fetch, you have to practice a little every day.

    Although it has been raining, I still want to ride my bike to school.

Got the idea? Use a caret to insert commas in the following three sentences.

1. Once you put down your homework it's hard to pick it up again.

2. When someone writes you a note or email be sure you respond.

3. If I pass the next math test I'm going to celebrate.

Rule 2

- Use commas to separate items (words, clauses, or phrases) in a series.

    Examples

    On my taco I like beans, cheese, lettuce, and plenty of hot sauce.

    Next summer I plan to visit my uncle's farm, play with my dog, read at least three books, and run five miles a day.

Got the idea? Then use carets to insert commas into the following sentences.

1. My favorite subjects to study are math art music and drama.

2. Of all the animals at the zoo, the most interesting are the gorillas cheetahs kangaroos and giraffes.

3. If you can sing play an instrument dance or do an animal act you can be in our circus.

> **Rule 3**

- Never, EVER put a comma between the subject and verb.

    Incorrect Examples

    Ralph, shot the ball and scored a point. (That comma should NOT be there.)

    All five of us, love movies. (That comma should NOT be there either.)

Got the idea? Use delete marks to take out commas that do not belong:

1. Wanda, ran 14 miles; then she, leaped for joy.
2. The 16 puppies, tore through the house.
3. Geography, is my hardest subject.

Share and Compare

Compare your editing with your partner's. Did you add commas in the same spots? Did you both delete ALL the commas in the third section? Review the sentences with your teacher and ask questions about anything that is unclear.

Reading, Editing, and Checking

Following is an example of student writing. This writer was a bit uncertain about comma usage. She added commas that don't belong and left out commas she needed.

Before editing, read the passage aloud to get a feel for it. Then, with your pencil in hand, read again. Insert commas that are needed, and delete commas that should not be there. If you get stuck, look back at Rule 1, Rule 2, and Rule 3.

Whenever we have an all-school party my friends and I end up getting mad. There's always lots of good food music and games. Plus all of my friends, show up. Then the DJ, plays a slow song. As soon as the slow song starts the boys I'd like to dance with run from the room like it's on fire. Most of the boys, hate slow music. They could be shy self-conscious or a little unsure about their dancing skills. It, is hard to say. When they play a fast song everyone dances together in a big group. That way no one feels in the spotlight. I, love to group dance. My favorites are group dancing line dancing and Irish folk dancing. At the last dance I kept my eye on Jeron Drew and Christopher. I, kept hoping one of them would ask me to dance. Sorry to say not one of them, did. When the next dance comes around I, think I'll ask one of them! That, could be funny!

Share and Compare

Compare your editing with your partner's. Did you add and delete the same commas? Coach your teacher as he or she goes through the passage with you. Ask questions about anything that is unclear.

A Writer's Questions

Not all writing handbooks agree on comma rules. Why do you think this is? What is the primary reason for using punctuation anyway—is it about more than following rules? If you are making a very important point, would it be OK to bend the rules and put in an extra comma just to slow the reader down?

Editing Level 2: Presentation
A Recipe for Clarity

Do you like to cook? Even if you've never tried it, have you ever browsed through a cookbook just for fun? If so, you may have noticed that a good recipe is a masterpiece of organization. It has two main parts: (1) the big picture (which is vital if you do NOT cook and have never seen quiche Lorraine, for example) and (2) step-by-step instructions telling you just what to do. Good recipes have other important features, too, and you'll have a chance to make note of them in the next section.

A Warm-Up

Your teacher has collected a few recipes for you to look at. As you review each one, think about the key features that make this recipe style presentation work well. Are there some recipes you could follow yourself, even if you were all alone in the kitchen?

With your partner or in your writing circle, jot down some features that make at least some of the recipes especially easy to follow. Be thinking about how you could use those features in other sets of directions, too (not just recipes).

Features for Easy-to-Follow Directions

1. _____
2. _____
3. _____
4. _____
5. _____

You Be the Judge

The organizational structure of a recipe (big picture + steps) can work beautifully for any writing in which the purpose is to teach or provide directions. Look at the following two examples. Use what you learned in reviewing the recipes to give each example your best critique. With a partner or writing circle, offer specific suggestions for improvement. Make notes right on the copy to help you.

Example 1

Brushing Your Way to Healthy Teeth

Everyone knows that clean teeth are important and can lead to improved overall health. No one wants plaque or gingivitis. A healthy mouth means you will be able to eat the foods that are good for you, and you will feel like smiling when you pose for pictures. To have healthy teeth, you need to know how to brush. Follow these steps: 1. Brush properly at least twice a day. 2. Spend time on all your teeth, not just those big front ones. 3. Don't forget to brush your tongue, too. Taking the fuzz off your tongue will remove bacteria and keep your breath fresh. 4. Squeeze a pea-sized amount of toothpaste onto a soft-bristled toothbrush. When you brush, use short, up and down motions or back-and-forth motions. 5. Rinse well, with water, dental rinse, or mouthwash.

Example of brushing

Our group recommends these changes:

1. _____
2. _____
3. _____
4. _____

Example 2

Straight Down the Fairway—Easy As 1, 2, 3

You've got all that beautiful, **carefully maintained grass** stretching out in front of you as you prepare to tee off. But that's the last you will see of it, because you just hit your ball into the trees, again. It might be impossible to hit the ball straight every single time, but by following a few easy steps, you can drastically reduce the number of times you have to go searching for your ball in the woods or **chin-high grass**. If you want to become that golfer who spends most of the time in **manicured short grass**, then follow these simple tips.

Step One
- Maintain a good posture
- Upright stance
- Left shoulder up, right shoulder down
- Don't stoop
- Can you see the ball? Where are you looking?

Steps Two, Three, and Four

To make sure that your beginning position is similar to where you want to be when you strike the ball, make sure your RIGHT ARM is touching your STOMACH on the right side when you are in your upright stance. Think about your swing plane—make sure you are going for your TARGET in a straight line. Tip: Turn your HEAD to the right, maybe two or three degrees.

Step 5

See the ball with your left. Be sure the club is in a good position. Concentrate on a smooth takeaway and backswing. Take your normal swing.

You should be amazed by the results. You have just removed most of your back-country adventuring from your round. Fore!

Some Important Things to Know About Golf

Our group recommends these changes:

1. _____

2. _____

3. _____

4. _____

Share and Compare

Discuss your critiques with the whole class. Were the directions on tooth brushing easy to follow? How about the directions for improving your golf game?

A Writer's Questions

As you may know, some people are visual learners, some are auditory (listening), others are more kinesthetic (hands-on), and there are many who learn best with a combination of approaches. What works best for you? How would knowing the learning styles of your audience help you with the kind of presentation you just worked on?

Presentation Matters

For this part of the lesson, you will work with your writing circle to create recipe-style instructions for doing a task or achieving a goal you know quite a lot about. Imagine that your article will be printed in the local paper or your school paper. It will be read by students your age as well as by adults.

Choose any how-to topic of your own or use our list to help you think of one. We recommend choosing your own topic, something you know so well that you will not need to do extensive research. Following are some suggestions in case you cannot think of a topic on your own.

★ Our own idea: _____
- Middle school survival guide
- Earning extra cash for the stuff you want
- Shooting a free throw
- Making the world's best pizza
- Making a good impression on someone you like or admire
- Performing any skateboard trick (or other physical feat)

In setting up your project, begin by deciding who will do what. Some tasks you can work on together, but you may wish to assign others to individuals. Maybe some team members are especially good at art or editing, for example. Your final directions should include all of the following key elements.

- A big picture summary paragraph—the project or goal in a nutshell
- Clear, manageable steps (try to limit the number to five)
- Concise writing that makes each step easy to follow and quick to do
- Appropriate graphics that let the reader see the project, step by step
- Good font and design choices that make reading easy

If you have computer access, by all means create your project online. Be sure to include a copyright (©) symbol to claim ownership of your own work. And give full credit for anything you borrow or sources you cite. Do not include copyrighted material without permission.

Organization

Sample Paper 7

Score for Organization _____

Worms Are Everywhere (and That's Good!)

If you compare what it takes to make an earthworm happy with what it takes to make a typical eighth grader happy, you will find they're basically identical. Food, water, oxygen, and decent temperature conditions—that's all an earthworm really needs or wants. Given these conditions, earthworms will flourish and even help you, not unlike eighth graders.

Earthworms really are *everywhere*. (Again, much like eighth graders.) In one acre of land there could be up to a million worms. Don't freak out, though. Most of them are underground doing their wormy work. And what exactly is that? They're constantly tilling, breaking up and moving the soil. As they tunnel down, search out, and eat organic plant matter, worms mix topsoil with subsoil. They create excellent fertilizer in the process. Their castings (also known as *excretions* or simply, *worm poop*) are very rich in nitrogen, soil microbes, phosphates and other nutrients, creating an outstanding fertilizer for plants and lawns. Gardeners revere worms; so do golf course maintenance people. Worm castings are clean and odorless, and can actually be created by composting kitchen and yard waste. (Read up on *vermicomposting,* and you could build your very own worm bin for composting and creating worm fertilizer. It's easy, organic, and cheap.)

Earthworms do need one more thing to flourish, and that is, of course, other earthworms. Some people believe that the reason you see so many earthworms after heavy rains is that they came to the surface to avoid drowning. Since they breathe through their skin, however, worms can live under water for a time; they have no lungs to fill with water. Actually, they come out to look for mates. When the ground is wet, they can get around so much more easily (increased mobility) while checking out the other worms.

The next time you look at your yard or a grass field at school, think about all that is going on just under the surface—worms munching on dead leaves or plant parts, composting, fertilizing, and moving the soil right beneath your feet, all the while looking forward to the next shower. And when it rains and the worms come out, tell them they're looking good, and give them some space. You go, worms!

Sources

Mary Appelhoff. *Worms Eat My Garbage: How to Set Up and Maintain a Worm Composting System.* Kalamazoo, MI: Flower Press, 1997.

Nancarrow, Loren and Taylor, Janet Hogan. *The Worm Book: The Complete Guide to Gardening and Composting with Worms.* Berkeley, CA: Ten Speed Press, 1998.

Payne, Binet. 1999. *The Worm Café, Mid-Scale Vermicomposting of Lunchroom Wastes.* Kalamazoo, MI: Flower Press, 1999.

http://yucky.discovery.com/flash/worm/pg000102.html

http://www.veggiegardeningtips.com/earthworm-castings/

Sample Paper 8
Score for Organization _____

Art and the World

What exactly is art? If you go to an art museum, you can see some pretty weird stuff. You can see some really neat stuff, too, like paintings from other cultures or from other times. Art can show how other people see the world and how people in other times saw the world. So what is art? One art expert said art is about experience—when you look at art, what do you feel or think about? So what if you were in a museum and saw a board painted bright blue, and it was titled "Blue." Would you think it was art? You might just think that if a board painted blue is art, then anything can be art. You might feel that if this artist got paid for the blue board, then you'd like to become an artist, too. Yet that is only one perspective.

Many people prefer art that looks like what it is supposed to be—a painting of a house, a person, or an animal. That is called realism. Other people in the world may prefer art that makes them think of other things. Like impressionism or even surrealism. That's what museums are for. To view art from around the world and from different historical times. Art to each person is a different thing and that has always been true since the earliest times. This helps us define it.

Architecture is yet another example of art. Frank Lloyd Wright was a famous architect. You can visit houses or buildings designed by him. Architecture is not like a painting (or a board painted blue) that you could get tired of after it hangs on a wall for a while. Designing and constructing a building takes far more time and is far more complex than regular art. It requires thought, which is a major part of art. So what is art exactly and how does it fit into the world? That is the question. It may take us centuries to answer it.

Sources

David G. Wilkins, David G. *The Collins Big Book of Art: From Cave Art to Pop Art.* 2005. New York, NY: Collins Design, 2005.

Editors of Phaidon Press. *The House Book (Architecture).* Published in London, UK by Phaidon Press in 2004.

http://library.thinkquest.org/J002045F/art_styles.htm

Organization

Name _____ Date _____

Revising Checklist for Organization

☐ My lead sets the stage and gets your attention! OR

☐ I should begin this way: _____.

☐ My ending wraps things up and leaves you thinking. OR

☐ I should end this way: _____.

☐ This writing is easy for readers to follow. They will NEVER feel lost.

☐ I stayed with one main message or story, beginning to end.

☐ I have a surprise or two—not everything is completely predictable.

☐ Details and events seem to come at just the right moment.

☐ I would describe my overall design this way:

    ☐ Main idea and details or support

    ☐ Chronological events

    ☐ Comparison-contrast

    ☐ Problem-solution

    ☐ Series of questions with answers

    ☐ Visual description from first impressions to subtle details

    ☐ Other _____

☐ I used paragraphs to show small shifts in the story or discussion.

☐ _____ rated my writing:

| 1 | 2 | 3 | 4 | 5 | 6 |

Note Good organization guides your reader as if you were shining a light on a dark path. Did you shine a light—or leave readers in the dark?

UNIT 3
Voice

Have you ever eaten something with no taste whatsoever? As you were chewing, maybe you thought to yourself, "Why does anyone *eat* this stuff?" Why indeed. Without flavor, what's the point? Guess what. Writing needs flavor, too—if it's going to appeal to anyone. Where does that flavor come from? After all, writers can't just reach for the spice rack. But they can choose intriguing topics, dig for the most interesting information available—and let their passion for the topic show in every line. When writers sound as if they care about the message, they get readers to care—and read on. That's the power of voice.

In this unit, you'll explore the power of Voice by

- defining this trait in your own words.
- using knowledge and strong feelings to create a confident voice.
- matching voice to audience.
- revising to give your own writing more voice.

Sample Paper 9

Score for Voice _____

A Gift for Giving

If I asked you to picture a grandmother in your mind, I'm not sure that image would even come close to resembling my grandmother Shirley. Picture a 67-year-old woman who still swims daily (and I'm talking laps, up and down the pool), paddles a kayak (yes, she wears a helmet), and rides in a rodeo parade, complete with cowboy hat, boots, and leather chaps. Shirley (she won't let me call her Grandma) taught me to quilt, bake delicious pies and fluffy biscuits, stitch up a deep cut (yes, she is a doctor), and "curse like a sailor," as she would say. The swearing instruction was not something she intended—it just happened. We drive around a lot together, and the road is the only place Shirley shows her temper. Be warned—drive too close to Shirley and expect to hear, "Hey! Don't tailgate me!"

Of all the things Shirley has taught me, the most important is that every day we are alive, and everything we do is a gift. A smile we give to a stranger, a song we sing, or even pointing out someone else's bad driving is a gift. Shirley taught me that gifts don't always come wrapped up—and sometimes they're random or accidental. Last year for my birthday, she taught me to paddle a kayak. Now, that is something we can do together. The year before, I helped deliver Ginger's puppies—Ginger is Shirley's yellow Labrador. Ginger had four puppies, and as each one was born, Shirley would hold it up and say, "Thank you Ginger—what a beautiful surprise!"

One time I was spending the night at Shirley's house, we stayed up late reading a mystery novel aloud to each other because it was so good, we couldn't put it down. She had these great voices for each character. When it was my turn to read I tried doing the same voices—they didn't sound as good, but Shirley said we made a great team.

Voice

Name _____ Date _____

Shirley tells me that someday I'm going to make a terrific grown-up, a fantastic mother, and an amazing whatever-I-decide-to-be. And you know, I think she's right. I guess you could call that a gift, too—we often aren't aware that we are receiving a gift until later. Every moment I've spent with Shirley, even cursing at drivers, has given me the gift of self-confidence and learning to believe in myself.

Sample Paper 10
Score for Voice _____

Maps

If I could change one thing, I would change road maps. They are hard to read. They do not show the world the way it is in real life. This can be frustrating and make navigating difficult.

Last summer, my family and I went on vacation to see our relatives in Nebraska. We rented a car. My dad didn't want to put a lot of miles on our car. We could have also rented a GPS. The rental car person even showed us one. My dad said no; he isn't much of a gadget guy. My parents took turns driving, and I got to be the navigator. They thought this would help me learn about geography and maps and problem solving. We live on the East Coast, so it was a pretty long trip.

I used an atlas with a new map for each state. You just look up the state when you come to it, or if you're on an interstate, use the U. S. map. I liked seeing where we were headed, but the real roads never match what is shown in the atlas. Sometimes it was hard to find the right road or to make sure we were going the right way. Sometimes we had to stop to ask for directions. My mom didn't mind, but my dad hated that part. A couple of times we got lost. That meant extra time so once we got to our motel after dark. The restaurant was closed, so we had snacks for dinner.

I sometimes wonder whether the people who make maps ever use them. It also made me think of what a hard job mapmaking must be. I don't think I would choose mapmaking as a career. Or being a navigator, either.

Voice

The WRITER... gets deeply involved with the topic.

So the READER...
_____
_____
_____

The WRITER... speaks in a natural, individual voice.

So the READER...
_____
_____
_____

The WRITER... speaks with confidence.

So the READER...
_____
_____
_____

The WRITER... thinks about the reader while writing.

So the READER...
_____
_____
_____

Voice

Name _____ Date _____

Lesson 3.1

A Defining Moment

In this exciting lesson, your entire future will be defined by the path you choose and by how courageously you act in the face of danger. Okay, we're exaggerating a bit. Your destiny may not hang in the balance, but this moment is still important and definitions are involved—definitions of *Voice,* that is. Your deep understanding of this trait will help you shape your writing to suit both audience and purpose, whether you're working on a personal narrative, fictional story, technical report, poem, or persuasive essay. One of the best ways to define voice is by listening to the voices of others—so, let's get started.

Begin with What You Know Now

You may already have a pretty good understanding of Voice. Reflect for a minute about how you would define this trait in your own words. Use the writing space below to share your thoughts, even if it feels like a guess. (Don't look at any rubric or checklist—write from your own thinking.)

_____
_____
_____
_____
_____

Now let's see if we can stretch your understanding of Voice even further.

What Do You Hear?

Following are two very different examples of writing. Share each example aloud in your writing circle, thinking about what it makes you **see** and **feel**. Do you hear the writer's voice coming through? Use the scale following each passage to rate the writing from **1** (almost no voice at all) to **6** (a strong, clear voice). Then try to come up with several words to describe each writer's voice.

Voice 1

Historical fiction, professional writer

Longstreet felt a depression so profound it deadened him. Gazing back on that black hill above Gettysburg, that high lighted hill already speckled with fires among the gravestones, he smelled disaster like distant rain.

It was Longstreet's curse to see the thing clearly. He was a brilliant man who was slow in speech and slow to move and silent-faced as stone. He had not the power to convince. He sat on the horse, turning his mind away, willing it away as a gun barrel swivels, and then he thought of his children, powerless to stop that vision. It blossomed: a black picture. She stood in the doorway: *the boy is dead.* She didn't even say his name. She didn't even cry.

Longstreet took a long deep breath. In the winter the fever had come to Richmond. In a week they were dead. All within a week, all three. He saw the sweet faces: moment of enormous pain. The thing had pushed him out of his mind, insane, but no one knew it.

The Killer Angels
by Michael Shaara

Voice

Name _____ Date _____

How I'd rate this voice:

| 1 | 2 | 3 | 4 | 5 | 6 |

Almost no voice — *Strong, clear voice!*

Words that describe Voice 1:

_____

_____

_____

Voice 2

Informational piece, professional writer

Scientists estimate that we may have seen fewer than *half* of the visible creatures existing in the ocean. Think about it. The thousands upon thousands of creatures we know are only *half* of what awaits discovery in the ocean's depths?

Several hundred feet below the sunlit surface, as deep as the most experienced divers venture, we are still resting on the "top" of this liquid universe, floating in our comfort zone. As we head down, light disappears—so does warmth. Giant schools of tropical fish yield to bigger cousins: whale sharks, manta rays, and enormous jellies. Below 500 feet, we humans can no longer see. But the strange (to our eyes) creatures that live in these depths—eels and sharks, the occasional nautilus—have adapted well and can see us quite perfectly if we descend that far.

Deeper still, below 1,500 feet, we'd find animals that glow in the dark—producing their own light through a process called *bioluminescence*. It's an ingenious method of communication, rather like sporting your own neon lights.

Voice

Name _____ Date _____

> Surely, you might think, at the deepest known part of the ocean—down nearly 36 *thousand* feet, more than *six miles*—nothing, *nothing* could live. But it could. It does. Shrimp, flatfish, worms, and bacteria call the deepest, most profoundly dark canyons of the ocean home. Only one vessel, the remarkable little *Trieste*, has ever gone so deep.
>
> From an unpublished informational essay

How I'd rate this voice:

1	2	3	4	5	6
Almost no voice					Strong, clear voice!

Words that describe Voice 2:

_____
_____

Share and Compare

Meet with a partner to share your ratings and thoughts about the voice in each example. Share your honest response—even if it's totally different from your writing partner's. Most voices speak to some readers more than others. Be sure to add any new words you come up with to describe each voice.

Voice

Name _____ Date _____

A Circle of Voices

Now it's time to share a voice you chose yourself. Review it one more time to make sure you like your choice—and feel ready to read it aloud. Then meet with your writing circle to share and rate voices. Take turns reading aloud. Record your group's average response—you can estimate it. Have everyone (including you!) brainstorm words to describe the voice you share.

Author's name: _____

Title of the work: _____

How my writing circle rated this voice:

1	2	3	4	5	6
Almost no voice					Strong, clear voice!

Words we thought of to describe this voice:

_____
_____

The Defining Moment

Look back to the beginning of the lesson, where you shared your first thoughts about the trait of Voice. Think about the voices you have heard since then, what you recall about each one, and the words you and others found to describe them. Then reshape your personal definition, making any changes or additions you wish.

Voice is . . .

_____
_____
_____

Unit 3 • Lesson 3.1   101

Voice

Name .... Date ....

Share and Compare

Share your definitions in your writing circles—or with your whole class (your teacher's choice). Listen carefully to each definition. Did anyone come up with a way of thinking about Voice that you really like? Add it to your own definition, if you wish.

A Writer's Questions

How would you describe your own writing voice? Do you like it? If you could sound like one of the writers you've heard in the past couple of days, who would that be?

Putting It to the Test

Sometimes, in a writing assessment, you're allowed to have a rubric or a checklist to guide you as you work. If that's so, is it still important to have a personal definition of Voice in mind? Why?

102   Unit 3 • Lesson 3.1

Voice

Lesson 3.2

Writing with Confidence

Informational writing is based on research, experience, or analysis. What voice does this type of writing need? We're going to call it the *voice of confidence*—confidence that comes from knowing a topic well and caring about it deeply. Writers who do not know their topics well tend to write in a general way: *Elephants are interesting animals.* Or they write in a timid, inhibited way: *It's widely believed that elephants could at times be rather intelligent, at least in some circumstances.* This voice is not confidence inspiring. Readers find themselves thinking, "Does this writer really know anything about elephants?" As a writer, you want your readers to believe in you. Know your topic, care about it, share your thinking honestly, and the voice of confidence will shine through in every line.

Sharing an Example: *The Endurance*

Following is a passage from Caroline Alexander's book *The Endurance: Shackleton's Legendary Antarctic Expedition.* This book offers a true account of the 1914 attempt by Ernest Shackleton and his crew to cross the Antarctic continent on foot. They managed to sail to within 85 miles of their destination before their ship, the *Endurance*, became trapped in pack ice. This passage describes the ship's last moments, just before the ice closes in to crush it.

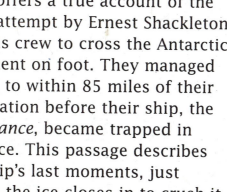

Unit 3 • Lesson 3.2 103

Voice

Read the passage aloud. Look and listen carefully for the writer's presence in the writing:

- How much does she care about her topic?
- Is she an expert?
- Does her voice inspire confidence?

The *Endurance* had quieted, but that evening an unsettling incident occurred while several sailors were on deck. A band of eight emperor penguins solemnly approached, an unusually large number to be traveling together. Intently regarding the ship for some moments, they threw back their heads and emitted an eerie, soulful cry.

"I myself must confess that I have never, either before or since, heard them make any sound similar to the sinister wailings they moaned that day," wrote Worsley. "I cannot explain the incident." It was as if the emperors had sung the ship's dirge. McLeod, the most superstitious of the seamen, turned to Macklin and said, "Do you hear that? We'll none of us get back to our homes again."

The Endurance: Shackleton's Legendary Antarctic Expedition
by Caroline Alexander

Thoughts and Reactions

Do you think the passage by Caroline Alexander has a strong voice? Use the writing space below to record two things this writer did to make her voice strong—or two things you wish she *would do* to make it stronger.

My Thoughts

Voice

Share and Compare

Meet with your writing circle to discuss your reactions to author Caroline Alexander's voice. What's your consensus? Is Ms. Alexander an expert—or does she have more research to do to inspire confidence? Did you and your writing circle identify similar strategies that this writer uses?

Reading, Rating, and Ranking

Read the following three informational writing examples to get a feeling for each writer's voice. Ask yourself:

- Is the writer deeply involved in this topic?
- Did he or she do any research—or take time to become an "expert"?
- Does the writer want me to care about this topic?
- Would I want to read more by this same writer?

Then rate each example from **1** to **6**, using this scale:

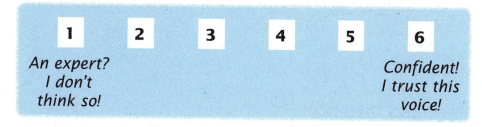

Example A

Snakes

Some snakes are really thin, and some are really big around. The biggest snakes come from the boa and python families, which include anacondas. Green anacondas can be really heavy, up to 400 pounds. These snakes move faster in water. Pythons, at 29 feet, are the longest snakes and can eat large animals, such as deer.

Rating _____

Unit 3 • Lesson 3.2 105

Voice

Example B

Conquering a Downed Power Line

You're driving with your family in the middle of a terrible storm. The wind is blowing tall trees back and forth in a wild dance, and pounding rain hammers the car too fast for the slapping windshield wipers to handle. Suddenly, a traffic light crashes to the ground, and the power line snakes across the hood of your car. Instinctively, you reach for the door handle—but should you step out? Remember: Whether or not that power line touching your car is sparking, you should *assume* it is alive with deadly electric current. Electricity can travel instantly through any conductive substance, including the rainwater outside your car. That first step out the door could very well be your last. Keep everyone calm, stay in the car, and use a cell phone (if available) to phone for emergency help. These actions could save your life.

Rating _____

Example C

Games and Violence

Video games and computer games aren't as bad as some people think. Some people think that kids who play violent games will go out into the world and be violent too. That will not happen! In fact, science shows that games have many benefits. People who play video games have better reflexes, better hand-eye coordination—and more fun! They develop important social skills. Kids who are playing video games are not out on the streets getting into trouble. Video games are helpful to society.

Rating _____

Voice

Name _____ Date _____

Ranking the Voices

Briefly review the ratings you gave each of the three voices. Then rank them, strongest to weakest:

Most Confident Voice _____

Middle of the Pack Voice _____

Least Confident Voice _____

Share and Compare

Meet with your writing circle to discuss your rankings. If you agreed, identify specifically what the top-ranked writer is doing to create a strong voice. How could each of the other writers revise to strengthen their voice?

The World's Leading Expert

Choose a topic you know quite a lot about—something on which you can sound like an expert for at least a paragraph or two. (You won't need to write a whole book!) You should also choose something you care about, so you can put your heart into the writing. Remember:
confidence = knowledge + engagement with the topic.

Choose any informational topic you wish—or use our list to help you think of an idea:

★ My own topic: _____

_____

OR . . .

- Playing any sport well
- Getting along with a difficult sibling (or other relative)
- Training or raising any kind of pet
- Growing a vegetable or flower garden
- Babysitting
- Secrets to good cooking
- Enjoying a specific kind of music or art

Voice

Once you choose your topic,
- take 5 minutes to prewrite, drawing a sketch or listing questions a reader might want you to answer.
- take another 15 minutes to write your draft, thinking to yourself, "I am the world's leading expert on this topic!"

Share and Compare

When you finish writing, share your draft in your writing circle. As others read their drafts, listen for confidence. Do you trust the writer? *Say so!* This is good—and encouraging—feedback to provide to your classmates.

A Writer's Question

What is the key to creating and maintaining voice in informational writing?

Putting It to the Test

Sometimes, in a writing assessment, you get stuck with a topic you don't like—and feel you don't know much about. Then how can you make your voice strong?

Lesson 3.3

Keeping Your Audience in Mind

Imagine you and some friends are headed to a concert headlined by your favorite musical group. The curtain rises, the spotlight hits the band, and to your amazement, the musicians begin playing . . . a *polka* (not that there's anything wrong with polkas). Are you disappointed, confused—even a little annoyed? Yes! It's the wrong sound, the wrong beat—the wrong *voice.* Like entertainers, writers need to know their audience if they want to be successful.

Sharing an Example: *Animal Farm*

Following is a short passage from George Orwell's classic novel *Animal Farm.* As you may know, this book tells the story of overworked, underfed, and otherwise mistreated animals who finally rebel, taking over the farm from their cruel master Jones. At first, the animals want to create a world based on justice and equality. But soon the pigs who instigated the rebellion are behaving a lot like Jones. Squealer, the pigs' spokesperson, has to find *just the right voice* to explain to the other animals why the pigs are getting better food.

Meet with your writing circle. Have at least two people read the following passage aloud. Listen carefully as each reader becomes Squealer. Does this crafty pig understand his purpose and audience well? Is he modifying his words—and voice—to get his way and manipulate the other animals?

Voice

"Comrades!" he cried. "You do not imagine, I hope, that we pigs are doing this in a spirit of selfishness and privilege? Many of us actually dislike milk and apples. I dislike them myself. Our sole object in taking these things is to preserve our health. Milk and apples (this has been proven by Science, comrades) contain substances absolutely necessary to the well-being of a pig. We pigs are brainworkers. The whole management and organisation of this farm depend on us. Day and night we are watching over your welfare. It is for *your* sake that we drink that milk and eat those apples. Do you know what would happen if we pigs failed in our duty? Jones would come back! Yes, Jones would come back! Surely, comrades," cried Squealer almost pleadingly, skipping from side to side and whisking his tail, "surely there is no one among you who wants to see Jones come back?"

Now if there was one thing that the animals were completely certain of, it was that they did not want Jones back. When it was put to them in this light, they had no more to say. The importance of keeping the pigs in good health was all too obvious.

Animal Farm
by George Orwell

Personal Reflection

Squealer wants the other animals to follow his thinking blindly. Think about how he uses his voice to get what he wants from his audience: *cooperation*. Can you think of two or three words to describe Squealer's voice? Write them here:

_____

_____

How do you picture the response of the other animals to Squealer's voice?

_____

_____

_____

Voice

Name _____ Date _____

Have you ever pulled a "Squealer"—using your *voice* to convince someone to follow your way of thinking? Describe the situation briefly.

_____
_____
_____

New Audience + New Purpose = New Voice

Squealer used a certain voice to speak to the other non-pig farm animals. Suppose for a moment that his **audience** had been his fellow ruling pigs—who are just as smart as he is—and his **purpose** was to convince them that sharing food with the horses might actually be a *smart* move. What sort of voice might he have used? Become Squealer for three or four sentences and speak to the ruling pigs. Make your voice convincing for this new audience.

_____
_____
_____
_____
_____
_____

Share and Compare

Meet with your writing circle. Take turns sharing your new Squealer voices. Are there any similarities? Since you're the writer, did Squealer wind up sounding a little like you? Or were you able to create a totally new voice for this character? Write two words that describe your "Squealer voice."

_____

Voice

Name _____ Date _____

Two Audiences, Two Voices

Part 1

Think of a personal experience you've had in the last year that you'll remember for a long, long time—one you'd share with relatives or friends. Take about two minutes to quickly record what you remember BEST about this experience—not whole sentences, just very quick notes:

_____

_____

Part 2

Now imagine telling the story of this experience to two VERY different audiences. You'll need to help each audience connect with your story. First, choose your readers:

1. **One adult** (parent or guardian, grandparent, other relative, teacher, coach, family friend, and so on) by the name of _____

2. **One person your own age** (close friend, sibling, classmate, teammate, cousin, and so on) by the name of _____

Part 3

Write two versions of your story—one for each reader. Think about the voice that will help that reader connect with your story. Remember, both details and word choice affect your voice.

Use your own paper. Spend about 8–10 minutes per story. Your teacher will tell you when half of your writing time is up.

Share and Compare

Meet with a partner or in a writing circle and share ONE of your stories—whichever version you like. Don't reveal your audience. See if your partner or writing circle can guess. And if they can, ask how they knew. Was it details? Specific words? Or were there other clues?

Voice

Name _____ Date _____

A Writer's Questions

We all use different speaking voices with different audiences: grandparents, siblings, teachers, police officers, infants, friends, pets, strangers. Do the ways you shift your speaking voice match the ways you shift your writing voice? How?

Putting It to the Test

In an on-demand writing situation, you do not really know who your audience is—specifically. You know it's an adult, and that's about it. Would it help to form a picture of this reader in your mind? And if so, what sort of picture would it be?

Lesson 3.4

Kicking It into High Gear

Writing *fueled* by a *powerful voice* takes readers on a wild ride with an emotional jolt. In contrast, writing that sputters along on the *fumes* of a *weak voice* leaves readers stranded on the side of the road, feeling abandoned. Here's the good news, though: with strategic revision, you can kick flat writing right into high gear. At the end of this lesson, you'll "refuel" one of your own drafts—just to see how high you can drive the voice.

Sharing an Example: "Rikki-tikki-tavi"

Voice is a formidable fuel. It can spark anger, fear, joy, amusement, tension, confidence—and a host of other feelings. Following is a short passage from the classic tale "Rikki-tikki-tavi" in *The Jungle Book* by Rudyard Kipling. Rikki-tikki is a mongoose, a creature known for its ability to kill snakes. He's been adopted by an English family living in India and protects them from Nag and Nagaina, a pair of dangerous cobras. This passage describes Rikki-tikki's battle with Nag. As you read, think about what you are **seeing** and **feeling**.

"It *must* be the head," he said at last: "the head above the hood. And when I am once there, I must not let go."

Then he *jumped*. The head was lying a little clear of the water jar, under the curve of it; and, as his teeth met, Rikki braced his back against the bulge of the red earthenware to hold down the head. This gave him just one second's purchase, and he made the

Voice

Name _____ Date _____

most of it. Then he was battered to and fro as a rat is shaken by a dog—to and fro on the floor, up and down, and around in great circles, but his eyes were red and he held on as the body cart-whipped over the floor, upsetting the tin dipper and the soap dish and the flesh brush, and banged against the tin side of the bath. As he held he closed his jaws tighter and tighter, for he was sure he would be banged to death, and, for the honor of his family, he preferred to be found with his teeth locked. He was dizzy, aching, and felt shaken to pieces when something went off like a thunderclap just behind him. A hot wind knocked him senseless and red fire singed his fur. The big man had been wakened by the noise, and had fired both barrels of a shotgun into Nag just behind the hood.

Rikki-tikki held on with his eyes shut, for now he was quite sure he was dead. But the head did not move, and the big man picked him up and said, "It's the mongoose again, Alice. The little chap has saved *our* lives now."

"Rikki-tikki-tavi" from *The Jungle Book*
by Rudyard Kipling

I can **see:**

_____
_____
_____
_____

I can **feel:**

_____
_____
_____
_____

Voice

What's the Secret?

With your partner or in your writing circle, look back at Mr. Kipling's version of "Rikki-tikki-tavi" one more time. As you read it together, put a letter *V* (for Voice, of course) in the right margin each time you think the voice comes through. When you finish, see if you can identify some specific things Mr. Kipling did to create voice in this passage. What are this writer's secrets? List as many as you can.

1. _____
2. _____
3. _____
4. _____
5. _____
6. _____

De-voicing

Now that you know what Mr. Kipling did to put voice in, it should be fairly easy to take the voice out! Just get rid of everything and anything that creates voice. It isn't every day you get to write badly on purpose, so have some fun. Work with a partner on this one and use scratch paper.

HINT: Your de-voiced version of the passage will probably be quite a lot shorter than the original. After all, you're taking things OUT!

Share and Compare

Meet with your writing circle to read your de-voiced versions of "Rikki-tikki-tavi." Talk about the specific things you did to take voice out of this passage. What's missing in your revision? What specific strategies had the most impact?

Voice

Refueling in Three Steps

Clearly, writing needs the "fuel" of voice to run well. Just how much fuel have *you* got right now? Let's see. Pull out your own rough draft—the one you plan to work on for this lesson. Then follow these three steps:

Step 1

Before making ANY changes, make a quick assessment: Just as you did with the Kipling piece, read through your rough draft and put a *V* in the right margin each time you think the voice is especially strong. Be honest. Underline any passages you feel have little or no voice.

Step 2

Share your draft aloud with a partner or your writing circle. Ask each person to offer one suggestion for making the voice stronger. Compare their comments against the marks you made on your writing.

Step 3

Revise. Use every trick you know:

- Review the strategies Mr. Kipling used to give his writing voice.
- Use the BEST strategies you got from your partner or writing circle.
- Picture your reader as you write. See the person in your mind.
- Say what you truly think and feel—and don't hold back.

Share and Compare

Meet again with a partner or in your writing circle—this time to share your revisions. Read your revised draft aloud, but also discuss your revision process. What helped you most? What strategies were most effective? What reader(s) did you picture as you wrote?

Voice

Name _____ Date _____

A Writer's Questions
Is voice a matter of attitude, or is it more about strategies? Is it both? Explain.

Putting It to the Test
In an on-demand writing situation, you don't—usually—get to choose your topic or your audience. So—what are three good tips for making sure anything you write in that situation still has voice?

Conventions and Presentation
Editing Level 1: Conventions
Real Speech

When people talk—or email—they don't necessarily follow all conventional rules. For example, because conversation is informal, speakers tend to use some slang. They also speak in single words and phrases instead of complete sentences, shorten words with contractions, or slide over pronunciations. (In writing we change spelling to show this.) Capturing the sound, rhythm, and flow of real speech in writing requires bending a few rules of conventions in order to make the dialogue sound authentic.

Strange as it sounds, though, bending the rules can take as much concentration as following them. It requires knowing rules SO well that you can manipulate them purposely to give your writing just the sound you want—like a jazz pianist running the scales. Authentic speech makes characters seem real—and that gives writing voice.

A Warm-Up

Part 1

Following are two sentences written in informal speech to reflect how a person—or a character in a book—might *really* sound. Read each one aloud and figure out what the person is saying. Then rewrite the sentence, using correct spelling, grammar, and punctuation. Be sure your revision is a conventionally correct, complete sentence.

Unit 3 • Conventions & Presentation

1. "Wouldja hold on a sec? I gotta git somethin'."
   _____
   _____

2. "Ya wanna come with? 'S gonna be great!"
   _____
   _____

Part 2

Here's a slightly different challenge—courtesy of ye olde days and William Shakespeare. It's actually an insult, taken from the play *All's Well that Ends Well*. First translate it into a grammatically complete sentence that reflects 21st century formal English. Then translate it into "real" speech—what you or a friend might actually say to someone if you meant to convey these very same ideas. **Remember:** Play with the language, but do not change the meaning!

3. "He's a most notable coward, an infinite and endless liar, an hourly promise breaker, the owner of not one good quality."

Updated Formal Version:
_____
_____
_____

Updated "Real" Speech (what a modern teenager might really say):
_____
_____
_____

Share and Compare

Meet with a partner to share your work. Did you translate the sentences in similar ways—even if you used different language? Try reading them aloud. Which ones sound the most like real speech to your ear?

Keeping It Real

Let's look at four bendable rules that can transform the sound and rhythm of "real" speech. You'll have a chance to warm up with each one—then use any or all of them to turn a very formal passage into something more authentic. Ready? Then let us venture toward reality. Or, uh, let's get real.

Guide to Authenticity

Rule bender #1: *Use contractions*

Before: I am going shopping at the mall. I will see you later.

After: I'm going shopping at the mall. I'll see you later.

Before: I would have another burger, but I am stuffed.

After: _____

_____

Rule bender #2: *Play with spelling to mimic <u>actual</u> pronunciation (using apostrophes to show missing letters)*

Before: I am going shopping at the mall. I will see you later.

After: I'm goin' shoppin' at the mall. See ya later!

Before: You are starting a fire under a tree? Are you crazy?

After: _____

_____

Rule bender #3: *Use phrases or fragments in place of complete sentences*

Before: Yes, I am going shopping at the mall. I will see you later.

After: Yeah—goin' shoppin'! Later!

Before: I am taking four science classes right now. It is quite challenging.

After: _____

_____

Rule bender #4: *Use appropriate but informal slang*

Before: I am going to see a movie.

After: I'm gonna catch a flick.

Before: I feel in the mood to spend some time with my friends.

After: _____

_____

Now that you're warmed up, have some fun turning the following passage from stuffy and formal into something a little more authentic. Imagine it's a passage from a novel written for students your age. The author is older than you are—but wants the voice of the narrator (the person speaking in this novel) to sound about 14 years old. Use your **Guide to Authenticity** to see what you can do. Make any changes you like. You're the editor!

Driving With My Brother

Would you allow me to share some interesting information? As of last Tuesday, my older brother Josh became an officially licensed driver. If one were to ask my opinion, the person who granted him this license might not be fully sane.

Allow me to explain what I mean. That evening, my mother and father stated that for the sake of their personal convenience, Josh will henceforth transport me to and from my soccer practice. When they relayed this to me, my face took on a new expression. I said, "Are you making a joke?"

Perhaps you are thinking that my primary concern is for my own personal safety. You are correct. But try to envision the following in your mind. As we were heading home yesterday, Josh exceeded the speed limit by more than 30 miles per hour and illegally entered two intersections on a red light. I continually cried out, "What are you attempting to do? Do you wish to put us in mortal danger?"

He smiled and replied, "I am currently in control of this vehicle! And please keep all further opinions to yourself."

Share and Compare

Meet with a partner to take turns reading your new versions aloud. Underline any sentence your listeners identify as especially authentic. Which rule-bending strategies did you make the most use of? Did you come up with any of your own?

A Writer's Questions

Bending rules to create authentic speech can lend significant voice to writing. Are there times, though, when formality is important? How do you know when it's all right to play with the rules—and when formality counts?

Editing Level 2: Presentation
Personalized Greetings

Ever found yourself standing in front of a rack of greeting cards, reading one after another, looking for the one that is just right? If so, you understand the importance of presentation and its connection to voice. The perfect greeting card has to have the right message (right for you, right for the person who'll receive it), color, art, and voice. As much as possible, it needs to look and sound as if you had written it yourself. Hey, wait a minute—that's a pretty good idea!

A Warm-Up

Have you ever sent or received a greeting card? Can you recall the occasions? Look at some of the cards your teacher and classmates have collected to share. Working with a partner or writing circle, brainstorm three lists:

- Occasions for which people commonly send or receive greeting cards
- Categories of people greeting cards are typically designed for
- Features you think make a good greeting card

Occasions for Sending Greeting Cards

1. _____
2. _____
3. _____
4. _____
5. _____

6. _____
7. _____
8. _____

My own idea: _____

Groups of People Greeting Cards Are Designed For

1. _____
2. _____
3. _____
4. _____
5. _____
6. _____
7. _____
8. _____

Features of a Good Greeting Card

1. _____
2. _____
3. _____
4. _____
5. _____
6. _____

Share and Compare

Meet with your writing circle to share your lists. How similar were your lists? Add any additional ideas to your list that come up during sharing. Also discuss the questions on the next page.

- Which part of a greeting card do you look at first (not including the slot where money could go)?
- Do you like funny or serious cards?
- Do you ever save cards you receive?
- Do you prefer paper or electronic greeting cards?

Presentation Practice

You Shouldn't Have! (Really!)

Have you ever received a greeting card from a relative or a friend that just didn't seem to match you or the occasion? Maybe you received a birthday card with a joke you didn't find funny. Or you got a card covered with cat photos—when everyone knows you're a dog person. Well, *someone* approved those designs!

Imagine *you* design and write greeting cards for a major publisher of such cards. (Actually, many people do this for a living.) You've been asked to preview two greeting cards to see if each is a match for the receiver and the occasion. Look carefully at all elements—referring to the list you made earlier. Then decide whether to (1) keep the card as is, (2) reject the idea totally, or (3) revise the card slightly so it's a better fit with the intended audience.

Example 1

Intended audience/occasion: Child turning six

My Decision

_____ Keep and publish as is—this will be a best-seller!

_____ Reject—a bad idea! Wrong card for the audience/occasion!

_____ Keep, but revise as follows:

_____

_____

_____

Example 2

Intended audience/occasion: Student graduating from high school

My Decision

_____ Keep and publish as is—this will be a best-seller!

_____ Reject—a bad idea! Wrong card for the audience/occasion!

_____ Keep, but revise as follows:

_____

_____

_____

Example 3

Intended audience/occasion: Get well card for someone not seriously ill

(You complete the inside—words and graphics.)

Share and Compare

Meet with your writing circle to compare your comments for each greeting card example and the words and graphics you used to complete Example 3. Did you agree about keeping, rejecting, or revising Examples 1 and 2? What do you consider the best design or writing ideas?

A Writer's Questions

Do people sometimes send greeting cards instead of making a phone call or a personal visit? Why? Why do you think more people don't create their own greeting cards for all occasions?

Presentation Matters

For this part of the lesson, you'll design a greeting card from scratch. You will need to choose a *real* person as the receiver—friend, parent or guardian, grandparent, uncle, aunt, cousin, teacher, or anyone you know. You'll also choose the occasion.

Keeping your audience in mind, create the message, both cover and inside; and design the layout, including art or pictures, color, fonts—the whole package. Make sure your message and design perfectly match the audience and the occasion.

NOTE: Make the message your own, even if it contains some familiar phrases and language—such as "Happy Birthday!" or "Congratulations!" Design or describe the art, specifying whether you want black and white or color, drawings, paintings, or photos, and so on. Use the space on the next page to sketch your art ideas, if you want.

Greeting Card Details

1. **Receiver** (Who is the card for?):
   _____

2. **Occasion** (Why are you sending this card?):
   _____

3. **Cover message**: _____
   _____
   _____

4. **Inside message**: _____
   _____
   _____

5. **Color scheme**: _____
   _____

6. Graphics (pictures, art, and so on)

Cover: _____

_____

Inside: _____

_____

Score for Voice _____

Sample Paper 11
Score for Voice _____

Tornadoes

Tornadoes can be scary. They can form out of any kind of storm. It's really only a tornado if it comes into contact with the ground. If it doesn't come into contact with the ground, it is called a *funnel*. Waterspouts are tornadoes that are formed over water.

Falling hail is one of the big dangers of tornados. Hail the size of golf balls or bigger falls through the tornado's updraft, developing layer upon layer of ice. Hail this size can hurt people, cars, or crops. A woman from Dallas, Texas, reported how scared she was after reaching the safety of a tornado shelter. She was in her basement for three or four hours with only a flashlight and a bottle of water. And flash floods are often caused by all the rain that occurs before and after a tornado.

The Fujita Scale is a scale for rating tornadoes. It is named after a scientist who studied tornadoes. It's like the Richter scale for earthquakes. If you are in your car during a tornado, you shouldn't stay in your car if the tornado is close. You should get out of your car and look for some kind of solid shelter like a basement.

Texas, Oklahoma, Kansas, and Florida get the most tornadoes. Oklahoma City has been hit by the most tornadoes of any city. There is a tornado season, when tornadoes are more likely to happen, but they can form any time the weather conditions are just right.

Here are some signs that a tornado could be forming: a cloud like a wall, really dark sky, large hail, and a roaring sound like a train. We should all watch for these important signs.

Name _____ Date _____

Sources

Mathis, Nancy. *Storm Warning: The Story of a Killer Tornado.* New York, NY: Touchstone, 2008.

1997. Verkaik, Arjen and Jerrine. *Under the Whirlwind: Everything You Need to Know About Tornadoes But Didn't Know Who to Ask.* Elmwood, Ontario, Canada: Whirlwind Books.

Mike Hollingshead and Nguyen, Eric. **Adventures in Tornado Alley: The Storm Chasers**. New York, NY: Published by Thames & Hudson in 2008.

http://www.nssl.noaa.gov/edu/safety/tornadoguide.html

http://skydiary.com/kids/tornadoes.html

http://www.livescience.com/environment/090618-tornado-facts.html

Voice

Sample Paper 12
Score for Voice _____

Sibling Rivalry

Scene #1

Adam (13) and Haley (11) are brother and sister. Every morning, Haley screams at Adam to hurry up in the bathroom so she can have a turn. Adam tells her, "Get over it! I'm older—end of story." Haley screams for mom or dad—who pull out their hair.

Scene #2

At least twice each week, Adam catches Haley snooping in his room. He explodes: "Why are you here???!!! Get out!!" She screams loudly, "Ouch! Heeeeeelp! I hate you!" Adam yells for mom or dad—who pull out more of their hair.

Almost anyone who has grown up with a brother or a sister—or is a parent with more than one child—can relate to these scenes. As badly as parents want their children to love one another and get along *all day every day*, it almost never seems to happen. Are the children innately aggressive? Are the parents incompetent? Actually, it's just the nature of brothers and sisters to fight. Scientists even have a fancy name for it: *sibling rivalry*. It's all part of growing up with a brother or sister. In fact, some psychologists say that it's *healthy*—because (provided we survive) it teaches us to live with people and deal with problems. But—does it have to be so . . . loud? Stressful? Frequent? Maybe not.

Like almost any stressful situation, fighting between siblings intensifies when we are overly tired, hungry, bored, or stressed. Remember Scene #1? The Battle of the Bathroom? What's really happening here? Maybe Adam and Haley stayed up too late. Maybe they skipped dinner and now they're hungry as tigers. Haley is worried she'll miss the bus—while Adam is fretting over that math test he should have studied for but didn't.

Voice

These two can't avoid getting up and going to school, but there are ways to minimize the stress. First, they need a schedule. That starts (dull as this sounds) with appropriate bedtimes so they don't wake up tired. Then—here comes the innovative part—they wake up at *different times*, thereby avoiding the usual hallway collision. Adam, who's older, gets a choice—wake up first to shower, or sleep in while Haley does her thing. They hit the breakfast table at different times, too—where they're delighted to find their favorite foods (waffles for Adam, cereal for Haley) ready and waiting. The goal is to keep them from crossing paths during the morning "rush hour."

On to Scene #2—The Maddening Snoop. Why can't Haley mind her own business? Actually, there's a good reason. Most kids admit that when they aren't involved in something interesting, the potential for trouble soars. Haley most likely pesters Adam when she has nothing more exciting to do. It could be that (once she finishes her homework), Haley needs a hobby (an aquarium, photography) or a sport she finds engaging. She could also spend time with a friend, go online, walk the dog, or help make dinner. Adam could help by setting a time when it is okay for Haley to visit. Once Adam's room is no longer forbidden, it won't seem so mysterious and enticing. And of course, Haley will learn that *she too* has a right to privacy and can set boundaries of her own.

What about those parents? The ones we left pulling out their hair? They need to get involved, perhaps setting up a family meeting to discuss problems and negotiate solutions. That meeting needs to happen when everyone is rested and stress-free—homework done, no big tests coming up, no recent fights. It's also important to hear everyone's point of view (siblings are almost never as ridiculously unfair and vicious as they appear in the heat of battle).

Some sibling rivalry—and the parental hair loss that goes with it—may be inevitable. The trick is to take all the energy and imagination that goes into rivalry and channel it into solutions.

Voice

Name _____ Date _____

Sources

Faber, Adele. *Siblings Without Rivalry: How to Help Your Children Live Together So You Can Live Too.* New York, NY: Harper Paperbacks, published in 2004.

He Hit Me First by Louise Bates Ames. New York, NY: Warner Books, 1989.

Randi Morris. *Dealing With Sibling Rivalry.* Lifescript: Healthy Living for Women (Lifescript.com). June 6, 2007.

http://www.lifescript.com/Life/Family/Parenting/Dealing_With_Sibling_Rivalry.aspx?page=2

http://kidshealth.org/parent/emotions/feelings/sibling_rivalry.html

http://www.med.umich.edu/yourchild/topics/sibriv.htm

Voice

Name _____ Date _____

Revising Checklist for Voice

☐ I feel strongly about this topic, so it was EASY to show that. OR
  ☐ I plan to change my topic to _____

☐ I know a LOT about this topic, so I sound confident. OR
  ☐ I plan to get more information from _____

☐ I read this aloud to myself, and it sounds *just like* ME.

☐ _____ rated my writing for Voice:

1   2   3   4   5   6

☐ I think a reader would *love* to share this aloud.

☐ I have highlighted any parts that need to be stronger. I plan to
  ☐ add details to make the writing more interesting.
  ☐ say what I *really* think and feel (write as if I *mean it*).
  ☐ use different words to give the writing life or energy.

☐ This is my purpose: _____

  My voice is: ☐ a good fit for this purpose ☐ not quite right yet

☐ Here's how I want readers to feel: _____

☐ My voice will make them feel this way.

☐ I used punctuation carefully so that when someone reads this aloud, it will sound *just the way I intended it to sound*.

> **Note** Your voice is really YOU on the page. Are you there? Are you at home in your writing? Do you speak right *to* readers?

UNIT 4
Word Choice

Imagine that you're making your way through the cafeteria line when a friend suddenly grabs your plate and says, "Here—let me get some food for you!" Before you know it, you're looking at a hamburger, pepperoni pizza, and teriyaki chicken. You politely remind your friend, "I was kind of thinking about a salad, some rice, and fruit." She looks very puzzled, as if you've lost your wits. "Come on," is her response. "Food is food, isn't it?"

Well, not exactly, especially if you are a vegetarian! As you'll discover in this unit, good writers are every bit as careful about their word selection as a particular eater is about food. Words, like foods, have their own textures and flavors—and when you have something special in mind, nothing else will do.

In this unit, you will and apply practice thoughtful Word Choice by

- exploring effective uses of synonyms and antonyms.
- harnessing the power of sensory words.
- choosing words that convey a precise message.
- achieving balance by cutting clutter—without sacrificing detail.

Sample Paper 13

Score for Word Choice _____

Having Braces

Having braces is something a lot of people have to go through. If you have or have had braces, then you know it is not a great experience. I think that it's really, really unfair that people like my brother get away with not having braces at all. I don't see why he gets to be so special.

My dentist says that I will need to wear braces for at least two years if I do everything I'm supposed to. Two years is a really, really long time. Every few weeks (this is the really bad part) I have to go see my dentist so she can put new wires in. This is what some people call getting their braces "tightened." Getting new wires hurts. It's not really like the worst thing in the world or anything, but it is bad enough, trust me. It hurts enough so that it is hard not to think about how much your mouth hurts. I never feel like eating or talking, or even texting. I don't even feel like smiling. Right after I get new wires, I have a hard time concentrating on schoolwork. My parents won't let me use that as an excuse.

My parents tell me that I will be really, really thankful that I had braces later in life when my teeth are nice and straight. I am sure they are right. Who doesn't want straight teeth, right? They are usually right about things in the end. But right now I really wish my teeth were nice and straight right now. Then I would not need to have braces at all. That would make me so happy I might even smile!

Sample Paper 14

Score for Word Choice ____

Warning: for Adults Only

Did those words make you hesitate for a moment—or want to plunge right into my paper? Frankly, I think they're misleading. When I see movie or television announcements that say, "Warning: This program contains adult language," I always wonder what on earth the filmmakers or programmers are thinking. Do they really believe language that is unacceptable for *me* to use is perfectly fine for anyone over 18 or 21? Are they sending adolescents some subtle message that being "grown up" is defined by the right (and ability) to use foul language like there's no tomorrow? Is this what "mature" people believe? If so, I respectfully disagree.

The very first time I saw the so-called "adult" language message, I was very young. I remember thinking, "This is going to be really boring. All they are going to talk about is taxes, bills, wrinkles, and hair loss." Instead, the show was filled with what my grandmother would have called "gutter talk." Aren't adults supposed to express themselves in a mature, socially acceptable way? That's what my parents tell me, but apparently they've got it wrong. I guess being "grown up" means finding a way to squeeze every conceivable obscenity into your writing or speech.

If we were truly mature, we'd develop a preference for thought-provoking entertainment. We would want to hear language used well—not in cynical or demeaning ways. As television and movie consumers, we should demand to be enlightened and educated. A little optimism or encouragement wouldn't hurt, either. Must TV programs consistently broadcast horrifying, dangerous, and depressing stories?

Name _____ Date _____

Every now and then we might be interested in something more inspiring: tutoring nonreaders, providing medical aid to remote areas, rescuing endangered species, or discovering a breakthrough way to make food and water safe.

I know that by expressing these opinions I run the risk of being ridiculed by my peers. They've been conditioned to think of violence and foul language as entertainment—and filmmakers are only too happy to comply. Most programs these days don't require intelligent viewing. They don't call for any deep understanding or thoughtful interpretation—unless you count trying to figure out why people find them enjoyable. They don't stretch our imaginations or teach us words that are truly useful anywhere but on the street.

This could change. Someday, I want to see the "For adults only" label used to mean "Warning: This program will require you to think." That would be a truer, more respectful definition of maturity.

Word Choice

The WRITER... chooses words with just the right shade of meaning.

So the READER...
_____
_____
_____

The WRITER... uses sensory words and phrases.

So the READER...
_____
_____
_____

The WRITER... uses powerful verbs.

So the READER...
_____
_____
_____

The WRITER... keeps it concise.

So the READER...
_____
_____
_____

Lesson 4.1

The Right Shade of Meaning

Synonyms are words with similar meaning—but they are not identical. If they were identical, we wouldn't need them. We have them because each expresses its own shade of meaning. Take a word like *big*. It has numerous synonyms: *huge, gargantuan, enormous, bulky, towering, spacious, expansive, vast,* and *whopping*—to name just a few. These are not interchangeable. You can have a spacious living room, for example, but not a bulky one. Your appetite might be huge, but you wouldn't call it towering. Get the idea? Grab your thesaurus and dictionary, and let's explore more shades of meaning.

Sharing an Example: *Great Expectations*

In addition to *A Christmas Carol, Oliver Twist,* and *A Tale of Two Cities,* Charles Dickens wrote this bildungsroman (a coming-of-age story) about an orphan named Pip's journey from boyhood to adulthood. In the following excerpt, Pip has been invited to the gloomy home of Miss Havisham to play cards with her adopted daughter Estella. Left at the altar in her youth, old Miss Havisham still wears the wedding dress from years before. As you read this passage, pay particular attention to the word *splendid*—a word on which Dickens puts particular emphasis.

Word Choice

Name _____ Date _____

> In an arm-chair, with an elbow resting on the table and her head leaning on that hand, sat the strangest lady I have ever seen, or shall ever see.
>
> She was dressed in rich materials—satins, and lace, and silks—all of white. Her shoes were white. And she had a long white veil hanging from her hair, and she had bridal flowers in her hair, but her hair was white. Some bright jewels sparkled on her neck and on her hands, and some other jewels lay sparkling on the table. Dresses, less splendid than the dress she wore, and half-packed trunks, were scattered about.
>
> I saw that everything within my view which ought to be white had been white long ago, and had lost its luster, and was faded and yellow.
>
> *Great Expectations*
> by Charles Dickens

Reflect

From the author's use of the word *splendid* and the context in which it is used, what do you think the word means?

Splendid probably means _____.

Check It Out

First go to your **dictionary** and look up the word *splendid*. Compare your definition with the dictionary's. Add anything to your definition to make it clear and memorable.

Word Choice

Now look up the same word in your **thesaurus**. Write down three or four **synonyms** (words with similar meaning). Put a star by the synonym that you feel best fits the author's intended meaning. Then write one **antonym** (word with opposite meaning).

Synonyms for *splendid* (used as an adjective):

1. _____ 2. _____

3. _____ 4. _____

One antonym for *splendid* is _____.

The Just-Right Shade of Meaning

Artists mix colors like forest green, moss green, or lime green, for example, to paint what they want viewers to see. As you already know, synonyms have slightly different shades of meaning, too. Read the following three sentences aloud:

- He was **angry** about the broken window.
- He was **fuming** about the broken window.
- He was **annoyed** about the broken window.

The underlined words are synonyms, but does each reveal the same level of feeling about the broken window? Imagine these words as colors and place them on the chart on the following page. Put the strongest, most-intense word on the top and the mildest, least-intense word on the bottom. See if you can add one synonym for each of the words and write it in the second blank.

Name   Date

Word	Synonym	Antonym
____	_____	_____
____	_____	_____
____	_____	_____

Opposites

Use your thesaurus to find antonyms for *angry*, *fuming*, and *annoyed*. Make sure each choice has the right level of intensity. Add these antonyms to the third column of the shade chart.

Share and Compare

Meet with your writing circles to discuss your shade charts. Did you place the synonyms at the same intensity level? Did you come up with the same synonyms and antonyms? If the words you chose were significantly different—so different that you disagree about them—use a dictionary to help you make a final choice.

Word Choice

The Right Shade

Following is another example passage. Notice that some words appear in blue—and are underlined. Some of those underlined words are the original words the author used. Others are not. Working with a partner or in a writing circle, follow these five steps:

1. Read the piece aloud as written. (Ignore the blanks for now.)

2. Figure out the part of speech for each underlined word.

3. Check out several synonyms—and if it helps, an antonym, too.

4. Choose the word you think has the right shade of meaning, and write it in the blank. (It may or may not match the underlined word.)

5. Read the result aloud to make sure you like the sound of it.

Every 4th of July, my family and I go _____ to the fairgrounds to watch the pretty _____ fireworks show. We bring a big _____ blanket to lay on. The loud _____ music starts the show as the pretty _____ fireworks flash _____ in the night sky. The show always ends with a large _____ finale that's really great _____.

When you finish, your teacher will share the author's original. Compare your choices to his. Do you agree with the author's word choices?

Word Choice

Playing with Opposites

Let's experiment with antonyms. Again, work with a partner or in a writing circle. Only this time, use your thesaurus to replace the author's words (in parentheses) with antonyms. Write those antonyms on the blank lines. Choose words that are totally opposite in meaning and intensity from the originals. **Hint:** Remember to keep the part of speech—noun, verb, adjective, adverb—the same as the word in blue so that each sentence makes sense. **Note:** If the noun *storm* in the last line no longer makes sense for your draft, feel free to change it.

The night sky was _____ (tempestuous).

_____ (Fierce) winds _____ (howled)

_____ (violently) around the windows and roof,

_____ (threatening) us as we _____

(huddled) inside, _____ (terrified). At last, like a

_____ (disgruntled) _____ (tyrant), the

storm _____ (thundered) away over the hills.

When you finish, read your new draft aloud. What happened to the mood and voice of the piece when you replaced each word with an antonym? What happened to the images you created in a reader's mind?

Unit 4 • Lesson 4.1   147

Word Choice

Name _____ Date _____

Using What You Know

During the introduction to this unit, you identified a piece of your own writing to work on. Pull out that piece of writing now. Using your dictionary and thesaurus, replace two or three of the words you underlined with synonyms or phrases that have just the right shade of meaning. Read the result aloud. Do you like what you hear?

A Writer's Questions
In this lesson you have been using a dictionary and thesaurus to expand the word bank in your brain. What other strategies can a writer use to stretch his or her vocabulary? Is it a good idea to combine strategies?

Putting It to the Test
In the very last part of this lesson, you revised a piece of writing by changing just two or three words. How long did it take you to do that? How could this strategy help you in an on-demand writing situation?

Word Choice

Lesson 4.2

Using All Your Senses

Sensory language—words and phrases that activate the senses of sight, hearing, smell, touch, and taste—gives readers a guided tour of the writer's world. Through sensory details, readers see what you see, hear what you hear, feel what you feel. Choose your sensory words carefully, and you strengthen the traits of Ideas and Voice, right along with Word Choice.

Sharing an Example: *The Fellowship of the Ring*

Consider the following passage from J.R.R. Tolkien's classic fantasy tale *The Fellowship of the Ring*, the first part of *The Lord of the Rings* trilogy originally published in 1954. You may be familiar with the books, movies, or both. In this example, the author uses sensory language to create a fantasy world. Read the passage with a pen or pencil in hand. <u>Underline</u> any sensory words or phrases you find.

At the south end of the greensward there was an opening. There the green floor ran on into the wood, and formed a wide space like a hall, roofed by the boughs of trees. Their great trunks ran like pillars down each side. In the middle there was a wood-fire blazing, and upon the tree-pillars torches with lights of gold and silver were burning steadily. The Elves sat round the fire upon the grass or upon the sawn rings of

Word Choice

old trunks. Some went to and fro bearing cups and pouring drink; others brought food on heaped plates and dishes.

"This is poor fare," they said to the hobbits; "for we are lodging in the greenwood far from our halls. If ever you are our guests at home, we will treat you better."

"It seems to me good enough for a birthday-party," said Frodo.

Pippin afterwards recalled little of either food or drink, for his mind was filled with the light upon the elf-faces, and the sound of voices so various and so beautiful that he felt in a waking dream. But he remembered that there was bread, surpassing the savour of a fair white loaf to one who is starving; and fruits sweet as wildberries and richer than the tended fruits of gardens; he drained a cup that was filled with a fragrant draught, cool as a clear fountain, golden as a summer afternoon.

The Fellowship of the Ring
by J.R.R. Tolkien

Your Response

Look carefully at the sensory words you underlined. Record some of them to fill out the following chart. We've provided a few to get you started, but you should find many more!

Word Choice

Name _____ Date _____

I see . . . *green floor, wide hall,* _____

I hear . . . *blazing fire, footsteps,* _____

I smell . . . *the fire, warm bread,* _____

I feel . . . *warmth of the fire,* _____

I taste . . . *bread,* _____

Reflection

Take a moment to reflect on the sensory details you noticed and recorded. Were you surprised by how many there were? Was there a particular sense that dominated this passage? Use this writing space to record your thoughts.

_____

_____

_____

Creating the Writer's World in Your Mind

This time, your teacher will share a passage aloud. It's from *Jane Eyre* by Charlotte Brontë. This example of realistic fiction centers on the life of its British narrator and title character Jane Eyre. The novel recounts her orphaned childhood, her schooling at Lowood School (where this scene takes place), her experiences as a governess and teacher, and finally her marriage to the mysterious Mr. Rochester. In this scene, Jane and her classmates are looking forward to having breakfast on a cold, wintry morning, having eaten so little the night before. They wake up at dawn, sit through several classes, and wait and wait for this moment to arrive.

This time you won't be able to look at the words. Just listen closely and let impressions form in your mind. Then talk with your writing group and list everything you see, hear, smell, feel, and taste.

Word Choice

Share and Compare

Compare sensory details with the class as a whole. Think about the world Brontë created for you. What is most vivid in your mind? Was there a particular sense that stood out in this passage? When you finish your discussion, your teacher will share the passage a second time. Did the world in your mind match the world in the writer's actual passage? Did you add any details of your own to that world?

Pulling Readers into YOUR World

Have you ever had to wait and wait and *wait*—as Jane and her classmates did in the passage you read earlier? Maybe you were waiting for a person or for something to happen. Where were you? What did you do as you waited? Was it a positive experience or a disappointing one, like Jane's? Take a thinking minute: Close your eyes and recall the scene in your mind, remembering how everything felt, sounded, smelled, tasted, and looked.

Then make some notes. Jot down the sensory details that are most vivid and important. Notice little details a reader might not think of. Record them as follows:

I feel . . . _____

I hear . . . _____

I smell . . . _____

I taste . . . _____

I see . . . _____

Look at your sensory chart carefully, and underline three to five details you think are especially interesting or unusual. Focusing on those details first, create a paragraph that helps readers live your experience of waiting. Invite readers into your world by sharing what you see, hear, taste, smell, or feel. Use your own paper to write your paragraph.

Share and Compare

Meet with your writing circle. Take turns reading your paragraphs aloud. Listen carefully for sensory details. After each writer shares, write the strongest sensory impression from his or her paragraph on an index card, fold it in half, and give it to the writer. Do NOT open any cards until *everyone* has shared.

A Writer's Questions

Picture yourself visiting an Internet site devoted to tropical vacations. The site has no photographs and uses no sensory language to describe the vacation destinations—only names of islands. Would that site be effective? In what other kinds of writing is sensory detail not just important, but vital?

Putting It to the Test

Everyone knows how important sensory detail is in descriptive writing. But suppose the prompt in an on-demand writing assignment asks you to write a persuasive or expository essay. Will sensory detail still matter? Why or why not?

Word Choice

Lesson 4.3

Getting Precise

When you hear the word *car*, what do you picture? Make a quick sketch in your mind—no need for paper. How big is it? What color? Old or new? Shiny or half rusted out? Now let's say you hear the words "cherry red, ground-hugging Italian sports car." Did those words change the image in your mind—even a little? If so, you know the power of precision. General, vague language leaves readers to fill in the blanks, making their own sketches. That's fine—if you want them to call on their imaginations. But if you want readers to see, hear, and feel what *you* see, hear, and feel, then only precise wording will do.

First Sketches

Following are two short writing examples, each depicting a scene. Read each one aloud, softly, to yourself. Then use scratch paper to create a sketch of what you see in your mind. Your sketch can be

- a drawing.
- a quick list of details.

Example 1

I love the desert where I grew up. There were trees, interesting creatures—and amazingly enough, even some water.

Example 2

The ocean is different from everywhere else on Earth. Even lying in bed, you can hear and smell things you don't notice anywhere else.

Share and Compare

Share your personal responses and mental sketches with those of a partner. Did you have the same impressions? Discuss the passages. How much work did you have to do as a reader—and how much did each writer do for you?

Check the response that matches how you felt as you created your own sketch of each passage:

- ☐ I was inspired. Each writer helped me see the scene perfectly!
- ☐ The descriptions were helpful, though I had to invent a few details.
- ☐ I had to make almost everything up myself! The writers hardly told me a thing!

Reviewing Examples: *The Secret Knowledge of Water* and *The Hungry Ocean*

Following are the actual nonfiction examples on which we based our previous very-sketchy descriptions. As you will see, in describing their real-life experiences in the desert and on the ocean, these two writers are considerably more precise than we were. Carefully read each passage aloud. Underline the words or phrases you think are particularly precise or vivid—the words that help you make a detailed sketch in your mind.

Unit 4 • Lesson 4.3 155

Word Choice

Name _____ Date _____

Example 1

An early memory of the low Sonoran Desert where I was born is of my mother walking me out on a trail. I remember three things, each a snapshot without motion or sound. The first is lush, green cottonwood trees billowing like clouds against the stark backdrop of cliffs and boulders. The second is tadpoles worrying the mud in a water hole just about dry. Each tadpole, like the eye of a raven, waited black and moist against the sun. The third is water streaming over carved rock into a pool clear as window glass. These three images are what defined the desert for me. At an early age it was obvious to me that water was the element of consequence, the root of everything out here. Even to say the word *Sonoran* required my lips to form as if I were about to take a drink, and the tone of the word hovered in the air the same as *agua* or *water*.

The Secret Knowledge of Water
by Craig Childs

Example 2

I woke up one morning, at the age of twelve, to the smell of low tide. The scent of seaweed and tidal pools crept through my open bedroom window and tiptoed around the room, not overpowering, but arousing interest. Usually awakening to the faint smell of pine and the rush of wind in the trees, that day I was intrigued with the thick, musty odor of sun-baked salt and

Word Choice

Name _____ Date _____

mussel-covered rocks. My ears strained to pick up the slight sloshing of the tide as it swept in and out around the low-water-mark rocks and ledges. It seemed strange that having been surrounded by the ocean my entire life, this was the first time I noticed the screeching of the gulls and the drone of a diesel-powered lobster boat nearby.

The Hungry Ocean
by Linda Greenlaw

Share and Compare

Meet with your writing circle to discuss the words and phrases you underlined. Did you notice the same precise, vivid, energetic language? Rate each of the passages on a scale from **1** to **6**.

The Secret Knowledge of Water

The Hungry Ocean

After rating both, choose one or two favorite words or expressions of your own from each passage and record them here. (Make your choices individually.)

Unit 4 • Lesson 4.3 157

Word Choice

Name _____ Date _____

Favorites from *The Secret Knowledge of Water*

1. _____
2. _____

Favorites from *The Hungry Ocean*

1. _____
2. _____

Verbs: The Engine of Writing

Verbs make writing both precise and energetic. Look at the passages one more time and identify two verbs from each that you feel are especially well-chosen.

Verbs from *The Secret Knowledge of Water*

1. _____
2. _____

Verbs from *The Hungry Ocean*

1. _____
2. _____

Quick Questions: If you could read a chapter from one of these books tonight, which book would you pick? Why?

_____
_____

Vague to Vivid = Verbs + Details

The wonderful thing about flat, dull language? It can be revised! And you're in the right place at the right time to do that. The following sentences are vague. Use energetic, precise details and strong verbs to bring each one to life. We've completed one to show you what we mean.
(**Hint:** Create a clear picture and make it move.)

Name _____ Date _____

Vague: The boy had fun in the sand.

Precise: *After the tide went out, Andrew scooped up bleached shells and gray driftwood sticks and assembled them into a fortress—then he sat back on the warm sand and waited for the ocean waves to attack.*

Vague: The food was unusual.

Precise: _____

Vague: A stranger approached.

Precise: _____

Vague: It was cold.

Precise: _____

Vague: She seemed upset.

Precise: _____

Vague: He ran a hard race.

Precise: _____

Share and Compare

Meet with your writing circle to share and discuss your revised sentences. Do you have favorites? Choose one or two to share aloud with the whole class. Then rate your own revisions here.

My revisions

☐ used details + strong verbs to make the writing precise.

☐ added some detail, but still made readers fill in a lot of blanks.

☐ were still way too vague. I needed more precise language!

Unit 4 • Lesson 4.3 159

A Precise Poem or Paragraph

It's time for you to create some precise, vivid writing of your own—a poem or paragraph. For this practice, focus on any place that holds strong memories for you. Come up with your own idea or use our list to jog your memory.

- Our old tree fort
- A room in my old elementary school
- Our first house
- The old barn
- The back yard
- Our street
- My favorite place to eat/hike/fish/hang out

Spend some time prewriting. Make a sketch, do a word web, list details—or whatever works for you.

When you feel ready to write, take 15 minutes or more to draft a poem or paragraph. Remember: Create a clear picture—and make it move.

Share and Compare

Before you meet with your writing circle, quietly read through your writing aloud. Are there any vague words or phrases you could replace? Do that now. Add details—or strengthen verbs. Then, with your group, take turns sharing. As you listen to each writer, jot down your favorite word or phrase on an index card, fold it, and hand it to the writer after he or she finishes sharing. Open cards when everyone is finished sharing.

Word Choice

Name _____ Date _____

A Writer's Questions

Some people prefer books without pictures. They say they like to make their own mental pictures of what the writer is describing. Do you feel like that, too? Does this put extra pressure on writers to be precise with word choice?

Putting It to the Test

Using precise language is partly a matter of avoiding vague language. What are some weak, general words you know will make your writing too vague? Brainstorm a list. How can you keep their use to a minimum?

Stop the Clutter!

You're having a dream . . . All the words in your writer's vocabulary are lined up like a team on the sidelines, anxiously awaiting the signal to be called into play. Suddenly it's time to write, they're at the ready, and they *all* want to get in the game. You try to blow your whistle . . . but nothing comes out! Here they come, ready or not, all the words you know, descending on the blank page, wave upon wave of language cluttering your writing, smothering ideas, overwhelming your readers . . . and then you wake up. The page is still blank. Your red pen is right there on the table where you left it! It's not too late. You can still . . . *stop the clutter!*

Sharing a First Draft

An eighth grade student was asked to write about a sports experience. Here's his first draft. Read it carefully aloud, discuss it with a partner, and mark your response:

> Summer football camp started yesterday. It's fun. You learn a whole lot of skills that are very useful in playing the game.
>
> I like most things about it. It lasts from now until August.

☐ It's too short and too vague. I didn't learn much.

☐ It's just right—a good balance of detail and concise writing.

☐ It's full of clutter! I could cut this in half and it would be way better.

Word Choice

Sharing a Revised Draft

This student was told to add more detail to his writing. He made an effort to do just that, and the following revised draft is the result. Share it aloud in your writing circle and discuss your response.

Summer football conditioning camp began yesterday, and it will last through the first full week of August. Football conditioning camp is organized and run by the local youth football association and the varsity head football coach of our local neighborhood high school. Camp goes from late afternoon until early evening or until it gets dark, which is about 8. The camp is for anybody, but you have to be going into grade three, four, five, six, seven, or eight. I am going into grade eight, and this is my fourth year of camp. You don't have to have any previous football experience, but it probably helps if you have had some experience with throwing, catching, blocking, and tackling. If you haven't had much experience, the camp's purpose is to teach about the basics of football—throwing, catching, blocking, tackling, and running. One fun thing about football conditioning camp is that it is one chance to hang out with lots of people you know, like your friends and people from your school or neighborhood. From the time school gets out until the first full week in August, I don't think about anything except football and going to my favorite thing, football conditioning camp.

Word Choice

Name _____ Date _____

Our Response

How would you rate this revised piece?

☐ It's still sketchy. I need a LOT more information.
☐ It's very balanced—plenty of detail but still concise.
☐ This writer went overboard! Where's my red pen???!!!

If you said the writer went overboard, we agree. Detail—as you know—is good, but this writer got carried away. How many words do you think this revision has?

A third draft is needed—and that's where you come in. Read the passage again, pen or pencil in hand. Look *and* listen for clutter, crossing out words, phrases, or sentences that are clouding or crowding the message. You may need to reword some parts to smooth the flow. Work independently.

Share and Compare

Compare your revision with a partner's. Who cut more? Were either of you downright ruthless? Not to worry—sometimes that's a good thing. (Don't be afraid to be equally ruthless with your own writing!)

One More Comparison

Now your teacher will share our revision—just as an example. Our final revision is 103 words long. As you compare your version to ours, put a check by the comparison that is closest to the truth.

☐ I slashed even more—mine is down to a few sentences.
☐ I cut about the same amount.
☐ I found less to cut, but I still like my revision.
☐ I found less to cut but decided I would cut more next time.

Putting It All Together

At the beginning of this unit you created a rough draft on a topic of your choice. Right now you may choose to work on that—or any other rough draft from your writing folder. Take a moment to find the draft you wish to work on.

Before you begin to revise, reflect for a moment on what it takes to make word choice strong. Each of these could be part of your revision. See if you can list four strategies for improving word choice:

1. _____
2. _____
3. _____
4. _____

You'll use these skills—and any others you think of as you write—to improve the word choice in your rough draft. Feel free to add, delete, or change anything. Use a dictionary or thesaurus to help you.

You have about 15 minutes for this revision. If you finish early, you should

- read your writing one more time to be sure you like the sound of it.
- make any last-minute changes that make it ready to share in a writing circle.
- offer coaching to anyone who needs help.

Share and Compare

Share your revised piece in your writing circle. Before you actually share your writing aloud, share your revision process: What did you do—and how did you know to do it? If you feel your piece needs further revision, ask for help from your circle.

Word Choice

Name _____ Date _____

A Writer's Question
Remember the second draft of the football camp paper? In that draft, the writer added a LOT of detail—to the point of filling his writing with clutter. But is there an advantage to writing a draft like that? Explain your answer.

Putting It to the Test
In an on-demand situation, what are some quick and easy things a writer can do to revise for word choice?

Conventions and Presentation
Editing Level 1: Conventions

All Is Well and Good

As a writer, you work hard. So you want to make sure each word you choose is working hard for you. That means selecting precise nouns and active verbs, not overdoing the modifiers, and using words correctly. That way, you won't wind up with sentences like these:

- I only needed ten more scents for my bus fare.
- Her bike ran over my cupcake, totally festooning it.

In this lesson, we're going to take a close-up look at two words that are commonly confused: *good* and *well*. Do you know when to use each? Let's find out.

A Warm-Up

The words *good* and *well* are not interchangeable, yet they are frequently confused, misused, and abused. See how well you know these two good words by filling in the blanks below.

1. Andrew did a _____ job at lacrosse practice last night.

2. The French roast coffee smells _____.

3. Maddie cooks very _____ for a fourth grader.

4. You really did _____ on that test!

Before checking with a partner, take a moment to write your own rules for using *good* and *well*. Don't try to sound like a grammar book. Just think logically about how you made your choices.

Use *good* when _____
_____.
Use *well* when _____
_____.

Share and Compare

Meet with a partner to compare your sentences and your rules for using *well* and *good*. Did you

☐ agree on uses of *well* and *good* in the four sentences?

☐ have similar rules for using each word?

Be ready to share your ideas with your teacher and class.

Refining Your Rules

Let's see if we can expand your understanding of the rules for using *well* and *good*. Answer each of the following five questions. Once you finish, see if you want to refine the rules you wrote earlier.

1. The word _____ is ALWAYS an *adjective*, which means it modifies a noun.

2. The word _____ is the ONLY one that can be used as an *adverb*, which means it modifies a verb.

3. The word _____ is NEVER an adverb.

4. The word _____ can be used to complete this sentence:

   I am feeling _____ today.

5. The word _____ is the one more commonly misused in everyday speech. Write an example of this misuse here: _____
   _____.

Share and Compare

Meet with a partner to compare and explain your responses. Did you each answer the questions with the same choices? Be ready to share your thoughts about what you learned with your teacher and class.

It Sure Feels Good to Do Well

Did you do pretty good on that practice? If that sentence made you cringe, we're betting you did very well! So—let's try something a little more challenging.

In this paragraph, the writer used the words *good* and *well* a number of times—sometimes correctly, sometimes not. Read the passage carefully, both silently and aloud. Correct any words that are misused. Draw a line through the incorrect words to delete them and use a caret to insert the correct word.

Spending the holidays with my grandparents did me a lot of good. We had a good time, and they helped me study for my upcoming math test. I feel well about it, and I'm hoping to do extremely good. My grandmother is a civil engineer, so she is extremely good at math, and she did so well coaching me! If I don't do good on that test, it's not her fault! Mr. Ruff, my algebra teacher, does very good at making algebra seem easier than it is. He is also good at identifying problems that will give us trouble. I feel good about this math test—I just hope I feel well enough Friday to even take the test. I ate too much of my grandmother's apple pie! (Did I mention, she does pretty good at baking, too?)

Share and Compare

Meet with a partner to compare your edits. Did you make the same corrections? Be ready to coach your teacher through his or her editing of the paragraph.

A Quick Reflection

Your teacher is going to do a mini-lesson on another grammatical problem he or she has noticed in your writing (or that of other students in your class). What do you think it might be? To make a guess, take any sample of writing from your folder and review it carefully. Do you notice

- ☐ any errors relating to word choice?
- ☐ passages about which you have questions?

Meet with your writing circle to continue your review together. See if you can identify one or more trouble spots or questions that a teacher could coach you on. Here are a few common problems many students have with words:

- Missing or wrong endings on words (-s, -ed, -ing)
- Subject and verb do not agree (We **was** late.)
- Shift in tense (We were skiing when we **spot** a friend.)
- Shift in person (I love swimming because **you** feel so free in the water.)
- Wrong preposition (Look in Part 2 **from** this book.)
- Problems with verbs (It **costed** too much. We **brang** more salad.)
- Misuse of who and whom (**Whom** is at the door? **Who** did they elect?)

Choose ONE problem you think your teacher might focus on today and be prepared to share it with the class—both problem and solution. Then follow along as your teacher leads the class in a mini lesson. **Note:** If you don't get to present your lesson today, fear not. There will be another opportunity coming your way soon!

A Writer's Questions
Where do grammar rules come from, anyway? Why should we bother to learn them?

Editing Level 2: Presentation
Words that Sell

We are bombarded by advertisements every day—in print and via television, radio, Internet, phone, and billboards. You're probably not surprised to hear that a lot of advertisers target YOU. Well—not you personally, but certainly people your age. Is it working? Can you think of an ad that got your attention recently? You probably have little interest in mortgage rates or insurance policies, but marketers are definitely thinking of you and your friends as they put together advertisements for electronics, jeans, casual shoes, and gaming devices.

Marketing, of course, is a specialty. Knowing your target audience is critical, but it's not enough. As we'll see in this lesson, marketers must also think about the questions potential buyers might have. Then they use good design elements—color, layout, illustrations, animation—to answer those questions effectively. Oh—and did we forget to mention? Word choice is one of those elements. Good ads rely on words that SELL.

A Warm-Up

Part 1

Take a few minutes to explore the advertisements that your teacher and class assembled together. Notice how the ads are worded and also how the advertisers use design to make certain words stand out. See if you and a partner can identify the features that make a successful ad work well. List them here.

1. _____
2. _____
3. _____
4. _____
5. _____
6. _____

Part 2

As you were reviewing the sample ads, did certain words catch your eye? Did you just ignore or slide over others? That's not accidental. Marketers know that some language speaks to certain groups and not to others. They use different words to target people ages 14, 24, 34—and 64.

Imagine that you're 25. (Can you do it?) And, precocious person that you are, you're in charge of advertising for a firm that makes backpacks for teenage consumers. Your first task is to zero in on the descriptive words that might get people ages 12 to 16 to pay attention and really, *really* want the backpack being advertised. (Vague, lifeless words need not apply for this job.)

Read through the following list aloud, with both product and audience in mind. Discuss the words with members of your writing circle and move any words that strike you as "keepers" to the first-cut list.

Brainstorm Words **First-Cut Keepers**

1. rugged
2. cozy
3. practical
4. economical
5. eco-friendly
6. fun
7. comfortable
8. ergonomic
9. adjustable
10. functional
11. cool
12. organic
13. masculine
14. high-tech
15. convenient

Final Five

Realistically, the advertisement only has room for five words. Two will appear in **HUGE, BOLD** print at the top. The other three will appear in a smaller bulleted list right under a photo of the backpack. Remembering your audience—and the features you want to emphasize—write your final choices below. Then give your backpack a name.

Final Cut: Top two words to appear at the top of your ad are:

1. _____
2. _____

Final Cut: Additional three words to appear right under the photo are:

1. _____

2. _____

3. _____

We're calling our backpack: _____

Making Waves with Words

Let's say you are in the market for an extreme outdoor adventure, something wild and memorable you can do with a group of friends. Something like white-water rafting. SPLASH! SCREAM!

If you were really going on such a trip, and you had the job of choosing the company your group would go with, what are the top three questions you would want an ad for white-water rafting to answer? Think for a moment. Then write them here:

My Top Three Important Rafting Questions:

1. _____?

2. _____?

3. _____?

Share and Compare

Meet with a partner to share your questions. Are they similar? Out of your six, which are the best three? Star them.

Now work with your partner to create a "Next Three Questions" list to get at additional information you might have forgotten in your excitement about rafting. Write them here, working together this time.

Our *Next Three* Important Rafting Questions:

1. _____?
2. _____?
3. _____?

Before You Buy

Following is an advertisement for a river rafting company—a guide service that takes customers on wild rafting adventures. Carefully look over the ad, reading like a prospective consumer—with your top questions in mind. Do their word choices provide easy, clear answers to those questions? Put a check (✓) by each question you feel is well-answered.

THRILLING!
Breathtaking scenery!
ENJOYABLE!

★ Inexpensive rates!
★ Appetizing snacks!
★ Radical!

1-800-555-5555

176   Unit 4 • Conventions & Presentation

My Thoughts

Which of the following best describes your response to this ad?

☐ The ad answered all my questions, and I'm ready to book my trip!

☐ I have a couple more questions, but the ad was helpful.

☐ This ad totally missed the boat. It didn't really tell me anything.

How many of your questions remain unanswered? _____

Share and Compare

Take a few moments to share your reaction with a partner. Did you agree—or did the ad appeal more to one of you than to the other? What suggestions could you make to improve this advertisement?

A Writer's Questions

Ads don't have very many words compared to, say, novels or textbooks. Is word choice as important in an advertisement, then, as it is in a novel? Why or why not?

Presentation Matters

For this last part of the lesson, you and your writing circle will select a product to sell and then design and write an advertisement to both attract and inform consumers. You should describe any visuals that will appear in your ad, but your main job is to choose words that will speak to your customers—and to create an overall design that will catch their attention and encourage them to buy your product. Follow the steps on this checklist:

- [ ] Choose a product—something you like and think you could sell. Give it a name.
- [ ] Think about your target audience: people your own age.
- [ ] Thinking like your customers, list the top five questions they will want answered.
- [ ] Brainstorm a list of words/phrases that describe your product, make it appealing, and answer consumers' questions.
- [ ] Create a mock-up of your ad, following procedures your teacher suggests.
- [ ] Prepare to present your product for critique by targeted consumers (in this case, other members of your class).

Word Choice

Sample Paper 15
Score for Word Choice _____

Salmonella Basics

Salmonella is not a small salmon or Cinderella's lesser-known sister. It is a bacteria carried by some animals, and can be present in eggs, soil, water, animal waste, and raw meats. A salmonella infection in humans can occur any time the bacteria is transmitted through foods or from contact with kitchen surfaces. It is one of the most common infections in the U.S., yet it's nearly one hundred percent preventable.

If you do become infected, you will usually notice symptoms within three days: persistent headache, fever, nausea, vomiting, stomach cramps, and diarrhea. Because many of these symptoms are also connected with strains of the flu, you will need to provide a waste sample for a clear diagnosis. The good news is that in most cases, the symptoms fade or disappear within a few days, given plenty of rest and liquids.

Though salmonella usually runs its course quickly, most people would prefer not to contract it in the first place. Here are some tips to keep you safe.

Prevention begins with extra care in the kitchen. Since the bacteria can live almost anywhere, it's critical to wash *all* fruits and vegetables (including those you peel)—along with your hands and all cooking surfaces. The heat from cooking will kill most of the bacteria—provided it's intense enough. That means that regardless of personal preferences, you should avoid raw or rare meats or fish. A cooking thermometer should be used routinely for roast beef, baked chicken, ham, or turkey; the surface may appear well done even when the inside is raw.

Cooks also need to take care to not cross-contaminate by using knives or surfaces for meats—and then preparing vegetables or fruits without washing the tools in between. It is safest to use separate tools and cutting surfaces for different types of food.

Personal hygiene is also critical. One of the best preventions is something everyone has been told to do since early childhood: wash your hands thoroughly after using the bathroom. In this case, *thoroughly* means a full 30 seconds of contact with foamy soap and *hot* water. A cold water rinse will not do the trick—and it's impossible to kill salmonella by wiping your hands on a towel. If you're in a big rush, just remember that salmonella will slow you down significantly. Take 30 seconds to keep yourself healthy.

You're now armed with the best weapon to defend yourself from salmonella poisoning: good information. Don't keep it to yourself. Spread your knowledge to help those around you stay well.

Sources

Hirschmann, Kris. *Parasites!—Salmonella.* Farmington Hills, MI: KidHaven Press, 2003.

Shotwell, Thomas K. *Superbugs: E.coli, Salmonella, Staphylococcus And More!: Does Super Farming Cause Super Infections?* Bridgeport, TX: Biontogeny Publications, 2009.

http://kidshealth.org/parent/infections/bacterial_viral/salmonellosis.html

http://www.medicinenet.com/salmonella/article.htm

Word Choice

Name _____ Date _____

Sample Paper 16
Score for Word Choice _____

Cultivating Tomatoes

Fresh, homegrown tomatoes cannot be defeated for flavor. Perhaps that is the underlying reason why most home gardeners approach the enterprise of cultivating tomatoes with euphoria and zest. Relishing a bounteous harvest can be laden with hazards, however. A few simple things can facilitate you to grow the ultimate prestigious tomatoes within existence.

The first thing to accomplish is choosing a vivacious tomato plant. Prior to selecting, you should know if the tomato types are determinate—meaning they will grow to a specified height then cease growing, or indeterminate—meaning that they will flourish as high as you authorize. The transplant you choose should be healthy, with luxuriant green foliage, and a robust and mighty stem. Prior to submerging them in soil, you should warm the ground by engulfing the bed in dark plastic for several days.

When you implant the tomato, dig a deep, cavernous hole. Inject some mulch and water into this crevasse. Insert the deciduous plant into the hole. Be sure to thrust it deep into the ground, working partially up the length of your arm. Promulgate additional mulch to envelop the roots. Bury the plant up to its top leaves. Then amplify the mulch and water as required.

Be sure to situate the tomato in a sunny location. Tomatoes possess a strong preference for full sun. Apply some fertilizer, but exercise caution. If you apply too much, the tomato plant itself will burgeon, but will not stimulate production of tomatoes. Continue watering the tomato plant during the duration of growing. Failure to water regularly may result in the plant's demise. Once the fruit begins to ripen, deter the water a bit to help sweeten the fruit.

Now you are informed with the complete secrets of cultivating exemplary home-produced tomatoes.

Sources

Joan Neilson, Joan. 1999. *The Great Tomato Book.* Berkeley, CA: Ten Speed Press, 1999.

Wilber, Charles. Austin, TX: *How to Grow World Record Tomatoes: A Guinness Champion Reveals His All-Organic Secrets.* Austin, TX: Acres U.S.A., 1998.

http://www.helpfulgardener.com/vegetable/2003/tomatoes.html

http://gardening.about.com/od/growingtips/tp/Tomato_Tips.htm

"Ten Mistakes to Avoid When Growing Tomatoes," by Steve Reiners, Associate Professor of Horticulture Sciences, Cornell University, July 27, 2004. (http://www.nysaes.cornell.edu/pubs/press/2004/040727tomato.html)

Word Choice

Name _____ Date _____

Revising Checklist for Word Choice

☐ I found some strong words or phrases to highlight. AND . . .

☐ I underlined words and phrases I need to revise.

☐ I found my OWN way to say things. I avoided tired, overused expressions.

_____ rated my writing for Word Choice:

| 1 | 2 | 3 | 4 | 5 | 6 |

☐ Three verbs that really work in my writing: _____, _____, and _____.

☐ I used sensory details to help readers experience _____ sights, _____ sounds, _____ feelings, _____ smells, _____ tastes.

☐ I crossed out clutter (words I did not need).

☐ I know the meaning of *every word I used.* OR, I need to look these words up:

_____, _____, _____.

☐ I replaced ALL general words like *nice, good, great,* or *wonderful* with specific, descriptive words that show an insider's understanding of this topic.

☐ If I used any NEW words, I made sure the meaning was clear.

☐ I spelled my words correctly so readers would know which words I meant.

> **Note** The words you choose make a bridge of meaning from you to your reader. Did you take time to make the BEST choices you could? Would readers learn any new words from your writing—or is there a phrase or two that might linger in their minds?

UNIT 5
Sentence Fluency

If you've ever done long-distance running, cycling, or long boarding, you know the importance of establishing a smooth, sustainable rhythm. Moving five feet, and then stopping, then five feet more and stopping again just doesn't work well if you ever hope to reach the finish line. Like racing, writing has its own rhythm. But while athletes feel the rhythm of the road in their heart and feet, writers mostly use their ears. They write, they listen, and they play with sentence length and word patterns until everything comes together just right. This unit is all about the rhythm of writing and what waits at the finish line: a clear message.

In this unit, you'll practice strategies for creating text that readers love to read. You will learn about

- blending variety with purposeful repetition.
- creating logical connections between ideas.
- identifying the secrets to fluent writing.
- putting everything together to create fluent text.

Sentence Fluency

Name _____ Date _____

Sample Paper 17
Score for Sentence Fluency _____

Hold the Garlic, Please!

Attention! Attention all home cooks and restaurant chefs! Is it really necessary to include garlic in *everything?* I realize this is America, of course—home of the Food Network where celebrity chefs hold court, dazzling average Americans with the power and magic of garlic. Everyone has the right to eat garlic at every meal if they wish, and most people do. But what about the rights of non-garlic-lovers like me?

Garlic is served to me (uninvited) in everything from grilling spice rubs to mashed potatoes to Caesar salad. It's even in ice cream! What's more, those of us who prefer not to indulge must breathe in secondhand garlic when we talk to those who have indulged. Mouthwash does not help. Brushing is a temporary fix. Strong breath mints offer a thin disguise. As soon as you consume it, garlic consumes you, oozing from every pore and vaporizing to form a protective yet invisible cloud that only other garlic lovers can penetrate without harm.

I'll be honest. For several years, I've been working to change my tastes. After all, since garlic is so popular, I might as well learn to like it, right? A person can only go against the tide for so long. Well, sorry garlic fans, but I don't think it's going to happen. Many years and countless breath fixes later, I am no closer to enjoying garlic-flavored chips, noodles, green beans, or friends. Still, I don't suppose garlic freaks want to give up the pleasures of the clove either. (You can smell it simmering in butter on the stove right now, can't you? I knew it.) So how about a compromise? It's all I ask. Omit garlic from a *few* things: grilled chicken, chocolate, scrambled eggs, yogurt, oatmeal, popcorn, chewing gum. And while you're at it, explore the wonders of *other* spices, such as rosemary, thyme, dill, basil, and sage. Give your garlic-numbed taste buds (along with your human buds) the thrill of a new taste experience. Hey, what was that? Did you hear those gasps? That was the people around you, breathing garlic-free air for the first time. Aaaaahhhh.

Sentence Fluency

Sample Paper 18
Score for Sentence Fluency _____

Moving

We had just were barely settled in California then when my dad got transferred to another location it was bam! just like that we had to move everyone but me was excited I did not like it one bit. The thing about moving there's a million things but one thing about moving is you have to pack that means everything you own down to the tiniest things like pencils and paper clips and bathroom tissue and underwear and socks if you're like me and you're not a detail person plus you never organize anything makes it an incredible pain. My mom told me label all the boxes I packed from my room I just wrote "Mike's Junk" on every single one of them a whole bunch of them I still haven't unpacked my mom reminds me every day to "Unpack your junk Mike" once you unpack it's home I don't want this to be home I miss California.

On top of everything else I had to leave my two best friends we had only known each other for two years it seemed like we had been friends all our lives people always say you will make new friends they're right that doesn't make up for the ones you leave behind. No way. The other thing that really bothered me is we were moving to a totally different place where we never even visited before we moved there Iowa is completely different way different from California it is different in other ways too it has extremely cold winters we don't even have clothes for that.

Now we have been here for two months it feels like a lot longer the days go so slow you can't believe it they just creep along at this snail pace I do not have one single really good friend in school yet I do not know if I will ever in a million years get used to it much less really get to like it makes me crazy having parents who like to move like it is really exciting to move give me a break moving is traumatic it's hard work I'll probably be scarred for life. Their only answer is they will say it builds character so what if it does I'd rather be back with my good friends that means a lot more to me right now than having character. The only good thing right now maybe my dad will get transferred again this time I am ready I can't wait to leave for another thing I am already packed.

Sentence Fluency

The WRITER...
creates smooth, rhythmic, fluid sentences.

So the READER...
_____
_____
_____

The WRITER...
uses variety or purposeful repetition.

So the READER...
_____
_____
_____

The WRITER...
creates long and short sentences—even occasional fragments.

So the READER...
_____
_____
_____

The WRITER...
begins each sentence in a meaningful way.

So the READER...
_____
_____
_____

Sentence Fluency

Name .... Date ....

Lesson 5.1

Don't Repeat Unless You Mean It!

Following are two short passages by different writers:

Example 1

He looked in the mirror when he left the house. He looked in the mirror when he got into the car. He looked in the mirror in the lobby of the hotel. And when he entered his room, before he turned to give the bellman a tip for carrying the luggage, he looked in the mirror once again.

Example 2

I have a hard time with computers. I like word processing, but I find other things difficult. I have trouble with spreadsheets or charts of any kind. I also have trouble with forms in general.

....

One of these writers used repetition on purpose, for emphasis. Which one? How do you know? A little repetition, used to make a point, can be highly effective. Most of the time, however, variety adds to fluency in much the same way that several different types of food can make a dinner party more interesting.

Sentence Fluency

Name _____ Date _____

Sharing an Example: *After Hamelin*

In *After Hamelin,* author Bill Richardson uses the legend of the Pied Piper as a springboard for a new, imaginative tale. This book tells the story of Penelope, who awakens on her eleventh birthday to discover she's lost her hearing. The Pied Piper, who rid her city of rats, was never paid the gold he'd been promised. As revenge, he uses his hypnotic music to lure the children of Hamelin away, but Penelope, who cannot hear his song, is left behind. Read the passage quietly to yourself, highlighting the first two or three words of each sentence.

I am a harper's daughter. In our house, for as long as I could remember, the thrum and ring of the harp had been as common a sound as the clatter of dishes or the slamming of a door. Everyone knew there was no harper finer than my father. Banquets, festivals, state occasions: none would be complete if the virtuoso Govan were not on hand to strike the harp.

His fame was widespread. Apprentices came from near and far to study with him. There was always a young man staying in our attic room. Sometimes, if they were homesick, they would talk in their sleep. I would wake in the night and hear them moaning sad-sounding words in Italian, Spanish, Welsh, Portuguese. Not even the Plague kept them away. They were willing to put up with rats gnawing their shoes and chewing their strings so they might learn to play, and also learn how to make the harps for which my father was so celebrated.

From far and near they came, with all their talent and all their yearning. But no one, no matter how gifted, was able to convince a harp to sing as true as Govan. And no one, no matter how diligently he worked, was able to make a harp with a voice as pure as one crafted by the master. The Maestro. That is what

Sentence Fluency

they called Govan. The Maestro. He could charm the music out of wood. No one, Govan least of all, could explain how he awakened the melody in balsam or beech or fir.

After Hamelin
by Bill Richardson

Reflection

Take a moment to read just the words you highlighted. How much variety did you notice in the beginnings of Bill Richardson's sentences?

☐ A great deal of variety

☐ Some variety

☐ Almost no variety

Now read the whole passage aloud. Do you hear any sentence repetition? Put an "R" in the margin where you notice repetition in the way sentences are formed. What point (or points) is the author trying to emphasize?

_____
_____
_____

Looking for Variety

What about your favorite writers? Do they vary their sentences a lot, a little, or scarcely at all? Let's find out. Look through one of your favorite books and identify a passage to share aloud with your writing circle. Look for a passage that illustrates the following.

- A lot of variety—many different sentence beginnings
- Repetition with a purpose, done to make a point

Sentence Fluency

Name _____ Date _____

Rehearse your passage. Then share it aloud with your circle. Tell them to comment on what they like and why they think you chose that particular passage. Choose one passage from your circle to share with the whole class.

A One-Sentence Warm-Up

How many different ways are there to write a given sentence?

☐ Two or three

☐ Half a dozen

☐ Any number, depending on the writer's skill and imagination

If you chose the third option, good job! Now we have a challenge for you. See how many different ways you can write this sentence in just two minutes. You can change the wording and the word order but not the main message. Use scratch paper. Read it aloud quietly to yourself to get your mind thinking. When your teacher says "Go," begin your revisions. Be creative!

My team hit and pitched the best it had all year and won the game, one to zero.

Revising a Whole Passage

Now that you've seen how much can be done with just one sentence, try your hand at a whole passage. Before you read or revise, underline or highlight the first three or four words of each sentence. Then read the passage aloud to get a sense of the message, as well as the rhythm and flow. Make any changes you think will improve sentence variety and overall fluency. You don't need to rewrite. We left room for revision on the page.

Unit 5 • Lesson 5.1   193

The Softball Championship Series

My softball team advanced to the championship series for the first time in school history. We had to travel to Sacramento, California, for the big tournament. My team was so excited to get to play for the title, even after we found out which teams were in our bracket. We had to win our first two games against teams who had won the tournament nine of the last ten years. We decided to use this as our motivation rather than let it discourage us.

My team's first game was against the current champion, a team from Texas. We really came through when it counted. My team hit and pitched the best it had all year and won the game, one to zero. We used this big win to propel us through our other games. My softball team won the rest of its games to earn the chance to play for the title. We fought hard in the championship, but we lost a tough game, six to one.

Sentence Fluency

Name _____ Date _____

Share and Compare

Meet with a partner or in a writing circle to share your revisions. Read slowly so you can compare each writer's sentence beginnings with your own. Check each of the following strategies used by you or anyone else in your circle.

- ☐ Changing a sentence beginning
- ☐ Flipping a sentence around to reverse the order
- ☐ Changing the wording
- ☐ Combining sentences
- ☐ Shortening a sentence, or breaking it into smaller sentences
- ☐ Deleting a sentence

Creating Variety

Step 1: As a final step in this lesson, you'll write and revise a fluent piece of your own. Fluency improves when a writer cares about the topic. So choose something that's on your mind right now or use the following list to help you think of an idea.

★ My topic _____
- No fair
- Now or never
- Seeing things differently
- Best friend ever

Step 2: Take five minutes for prewriting. You might complete any of the following.
- Make a sketch
- List readers' questions
- List details
- Make a word web

OR do anything that puts your thinking in motion.

Sentence Fluency

Step 3: Write nonstop for 15 minutes. As much as possible, try to begin each sentence in a slightly different way *unless* you are repeating for emphasis!

Share and Revise

Before sharing, read your own writing aloud, underlining or highlighting the first three or four words of each sentence. Then share your writing aloud with a partner, who should listen for repetition. Does his or her impression confirm what your eyes tell you, looking at the passages you marked?

After sharing, take five minutes for revision, changing any sentences that don't sound quite right. Trust your ears to tell you what works.

A Writer's Questions

It probably isn't natural for most writers to start *every* sentence differently, even though we asked you to try that in this lesson. Does every sentence need to begin differently? Or could beginning even some sentences differently improve fluency significantly?

Putting It to the Test

In on-demand writing, too much sentence similarity can have a deadly effect. Why might this be the case? What can you do to prevent it?

Sentence Fluency

Name _____ Date _____

Lesson 5.2

The Logical Flow of Ideas

If you had a written record of your thoughts, they might not seem connected. That's because your mind can jump around (in time too short to measure) from breakfast to the math exam to an upcoming weekend adventure. Until you share those thoughts—aloud or on paper—you don't need to connect them because you're the only one who has to understand them. Readers, however, depend on connections. They count on writing to be logical. That means that each sentence seems to flow right out of the preceding one and set up the one to follow. Sometimes logical connections are made with transitions (however, next, all the same, on the other hand), sometimes with pronouns (he, she, it, this, that), and sometimes with an example or explanation. In this lesson, we'll look at ways of making the flow logical.

Sharing an Example: *Dracula*

Following is a slightly revised passage from the classic horror novel *Dracula* by Bram Stoker. In this passage, Jonathan Harker is a guest in the castle of the eccentric Count Dracula. He is shaving one morning when he has a visit from his host. We have altered the writing a bit (with apologies to the author), removing transitional clues that show how ideas connect. Read the passage aloud, softly, inserting a question mark each time you find it difficult to make a connection.

"Good morning."

I had not seen him. The reflection of the mirror covered the whole room behind me. I cut myself shaving. I turned to the glass.

Sentence Fluency

The man was there. I could see him over my shoulder. There was no reflection.

I saw the cut bleed. I laid down the razor. The Count's eyes blazed with fury. He grabbed. I drew away. His hand touched the beads of the crucifix. There was a change. I could hardly believe my eyes.

Reflection

How did this passage sound as you read it aloud? Did it have a natural flow? Could you identify the main idea, or did you have to work hard to make the connections?

The author's main idea is _____

_____

- ☐ It was very simple to make connections. Each sentence led right into the next.
- ☐ It was challenging to make connections. I had to try hard to figure things out.
- ☐ It was impossible to make connections. Nothing went with anything else.

The Original (Thankfully!)

Here's the passage as Bram Stoker wrote it. The examples and explanations that give the piece its logical flow have been restored. Read this version aloud and share your thoughts.

"Good morning."

I started, for it amazed me that I had not seen him, since the reflection of the mirror covered the whole room behind me. In starting I had cut myself, but did not notice it at the moment.

Sentence Fluency

Name _____ Date _____

Having answered the Count's salutation, I turned to the glass again to see how I had been mistaken. This time there could be no error, for the man was close to me, and I could see him over my shoulder. But there was no reflection of him in the mirror!

I saw that the cut had bled a little, and the blood was trickling over my chin. I laid down the razor, turning as I did so half-round. When the Count saw the blood, his eyes blazed with a sort of fury, and he suddenly made a grab at my throat. I drew away, and his hand touched the string of beads around my neck which held the crucifix. It made an instant change in him, for the fury passed so quickly that I could hardly believe that it was ever there.

Dracula
by Bram Stoker

Reflection

How did the author's original passage sound as you read it aloud? Did it have an easy and logical flow? Do you have a different sense, at this point, of the writer's main idea? Share your thoughts here.

I would describe the author's main message this way:

_____
_____
_____

This time around,

☐ I understood the piece much better.

☐ I still had to try hard to figure things out.

☐ I still couldn't make sense of the message.

Sentence Fluency

Name _____ Date _____

Restoring the Connection

Following are three sets of disconnected sentences. Try to figure out how they connect logically, then rewrite them to make that connection clear. There are no right answers. It's fine to combine or expand sentences. Use transitional words, explanations, or any other connections you wish. Your logical solution may require one sentence or more.

Example 1

Fallen leaves have clogged the city's drainage system. The forecast calls for rain.

_____
_____
_____
_____

Example 2

Boris felt no one liked him. He practiced dancing in his room.

_____
_____
_____
_____

Example 3

Aquarium hobbyists are crazy about seahorses. Seahorses rarely survive in captivity. Seahorses are a threatened species.

_____
_____
_____
_____

Sentence Fluency

Name _____ Date _____

Share and Compare

Meet with a partner from your writing circle and share your revisions. Did you follow the same logical paths? If not, that's fine, so long as each solution makes sense. There's always more than one way to build connections.

A Bigger Task

As you have seen, building connections is primarily a matter of figuring out the main message. After practicing with small sentence sets, you're ready to build logic into a larger message, this time on forest fires. Read the following piece aloud carefully, with a pen or pencil in hand. Use any of the following strategies to improve the fluency and logical flow.

- Add transitional words or phrases
- Use pronouns carefully
- Change wording
- Combine sentences
- Add examples or explanations

It's that time of year. Smoke fills the sky. Wildfires dominate the local TV news. Young men and women are at the airport. Volunteers are joining hotshot crews. Hot shots fight fires on the front lines. Yesterday, seven fires were burning. Crews have been fighting the fires for many weeks. Helicopters use huge buckets used to dump water on the worst spots. Helicopters have been seen in the sky. Some crews are now returning. They will get some rest.

Unit 5 • Lesson 5.2 201

Sentence Fluency

Name _____ Date _____

Share and Compare

Before sharing, read your revision aloud, softly, one more time. What is the main message? Does the logical flow make that message clear?

_____
_____

When you're ready, meet with your writing circle to share your revisions aloud. Listen carefully to each version. Are they different? Choose one to share with the whole class.

Letting Your Logical Thinking Show

For the final step in this lesson, you'll write and revise a short passage, making sure it has both (1) variety and (2) a logical flow to make the message clear.

Step 1: Begin by choosing a topic—something you know well and can write about confidently. It's much easier to think logically when you know your topic! Choose something that's on your mind right now, or use our list to help you think of an idea.

★ My topic _____
- Why music and art should be part of our curriculum
- Easy ways for kids to go green
- A film everyone should see (or book everyone should read)
- The best thing about our neighborhood

Step 2: Take five minutes for prewriting. You might complete any of the following.
- Make a sketch
- List readers' questions
- List details
- Make a word web

OR do anything that puts your thinking in motion.

Step 3: Write nonstop for 15 minutes. Go back once or twice to reread what you have written, asking yourself, "Does it make sense? Does each sentence flow naturally out of the one before?"

Share and Revise

Read your passage one last time prior to sharing, and make any quick revisions you feel are needed. Then share your passage in a writing circle. Have listeners share any questions they have about how ideas connect. Write those questions down so you can recall them later if you revise.

A Writer's Questions

As you know, setting your writing aside for a day or more can make a big difference in revision. Do you think it could help you discover gaps in logic? Try it with the draft you wrote for this lesson. Set it aside for three days. Then look again. Does it look just the way you remembered it? Can you revise with more confidence and power after not seeing it for a while?

Putting It to the Test

On-demand essays frequently receive low scores simply because the readers cannot see how ideas connect. What can you do to make sure this doesn't happen with your writing?

Secrets to Fluency

Reading truly fluent writing can be such a pleasure that you can lose yourself in the text, never hearing the doorbell or the voices of people trying to get your attention. Has that ever happened to you? If so, you know that fluent writing can sweep you along in its rhythm the way a river carries a raft. How does it do that? In this lesson, you'll explore several fluent (and less fluent) passages to see if you can uncover some of the secrets to fluent writing.

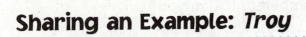

Sharing an Example: *Troy*

Following is a short passage from *Troy*, by Adèle Geras. The book is based on the story of the Trojan War, something you may have read about before. Read the passage aloud, softly. Think about some of the things this writer does to make this passage fluent.

Alastor opened his eyes and wondered for a moment if this was the kingdom of Hades. He was lying flat on his back on a thin pallet on the ground. This place was nowhere he recognized, but it smelled of men, sweating and bleeding, and there were sounds of groaning coming from somewhere. He tried to turn his head, but pain like knives heated on a blacksmith's fire and plunged into his neck stopped him, and he was left staring into the darkness about

Name _____ **Date** _____

his head. He closed his eyes. Had he really seen him again: the Black Warrior, the one who had come so close to him on the battlefield? He'd been there, yes, yes he had... in the corner of this very room, staring at me from under the iron helmet, his eyes as cold as death. I remember, Alastor thought, exactly where I saw the Warrior before. He said his name was Ares.

Troy
by Adèle Geras

Reflection

Rate the fluency of this passage from **1** (difficult to read) to **10** (like rafting the river).

| 1 | 2 | 3 | 4 | 5 | 6 | 7 | 8 | 9 | 10 |

Now see if you can list several specific things this writer did to make the fluency strong. These might be things you learned about in previous lessons or things you never thought about before.

1. _____
2. _____
3. _____
4. _____

Share and Compare

Meet with your writing circle to discuss your responses to the passage from *Troy*. Did you agree in your ratings? Did you notice different strategies for achieving fluency? At the beginning of this lesson, you identified a passage you felt was particularly fluent. With your teammates, share your passages now, listening for any additional strategies no one has mentioned yet. Finally, use your team lists to create a whole class list titled Secrets to Fluency. Post this where you can see it, refer to it, and add to it.

Sentence Fluency

Read and Rank

Reading is one of the best ways to improve your own writing. You stumble across problems to avoid and discover strategies you can try. Here are three pieces that vary in fluency. Read each example at least twice, aloud if possible. Ask yourself, "Is it easy to read? Does it flow smoothly?"

Example 1

I don't know if I could ever live somewhere that wasn't within at the most a two-hour drive from the beach. I like being close to the beach so that I can walk and collecting driftwood is fun with the sound of the surf washing away stressful thoughts, which helps me recharge my batteries, as my mom loves to say. I think the beach is the most special place on the planet. I want to write a novel one day I think I'll have the story set at the beach!

First thoughts: _____

_____

Example 2

This summer's most memorable experience was watching my brother learn to ride his bike. As the older sister, I feel responsible for him, even proud of him sometimes. Because he draws trouble like a magnet, we fight like cats and dogs. Still, I can't stand the idea he might really get hurt, which made it even scarier when he tried to show off for me. He'd been yelling for me to come outside and watch how fast he could go . . . and I watched, all right. He was looking right at me when he hit a bump. Before I could call out, he flew over the handlebars and hit the pavement. Hard. For a few seconds, everything went silent. I stopped breathing. Then I heard him call my name, and I ran.

First thoughts: _____

_____

Sentence Fluency

Name _____ Date _____

Example 3

   Even if not everyone agrees (and you probably don't) but I still think cats make the best pets or anyhow better than dogs or fish just thinking about all the pets a person could have. Even hamsters or rabbits. Because cats are independent. This means not needing a lot of attention or you won't need to talk to them all the time. Dogs are different. This is their nature. Also, cleaning the fish tanks. The other thing is you can leave your cat for a whole weekend it will be just fine when you return it might be a little upset with you. This is typical! Because, if you think about it, this is the most important thing.

First thoughts: _____

_____

Fluency Rankings

With a partner, discuss the three examples and rank them, using the guide below.

HINT: Keep the list of fluency secrets in mind, along with your first thoughts.

___ BEST IN FLUENCY: Top notch! Very easy to read aloud!

___ RUNNER-UP: Not bad, despite repetition and some bumpy spots.

___ NEEDS WORK: A challenge! Very little flow or logic!

Be prepared to share and defend your choices with the whole class.

Sentence Fluency

Name _____ Date _____

Revision Time

Working on your own, revise the example you think most needs work. Revise it by following these steps:

1. Read it aloud once more, softly, to get the main message.

2. Keep that message clearly in mind as you work.

3. Use scratch paper, or if you have computer access, do your revision digitally.

4. Revise the sentences one by one, using any strategy you wish.

5. Read aloud as you revise.

6. Leave big margins so you can make additional changes once you finish.

7. Do one final read-aloud after you finish revising.

Share and Compare

Meet with your writing circle to share your revision and compare it with other possibilities. Feel free to borrow additional ideas, fine-tuning your own revision.

Following your group discussion, watch and listen as your teacher models his or her revision of the same passage. Pay particularly close attention to the process your teacher uses.

Sentence Fluency

Name _____ Date _____

More Secrets

Think about all the different ways you have explored fluency in this lesson:

- Reading the example from *Troy* by Adèle Geras
- Sharing fluent examples you and your teammates found on your own
- Ranking three student examples for fluency
- Revising the weakest of those examples
- Watching your teacher model his or her revision

Have you uncovered any additional secrets to fluency? Discuss them, and add them to your list!

A Writer's Questions

As you revise your writing for fluency, what other traits come along for the ride? In other words, how else does your writing improve?

Putting It to the Test

What are some quick things a writer can do to check or revise for fluency in an on-demand writing situation?

Sentence Fluency

Name _____ **Date** _____

Lesson 5.4

Smooth Sailing

Whether piloting an airplane or an ocean-going ship, the captain is responsible for charting the smoothest course possible. A captain must do more than flick on the fasten-seatbelt sign or shout "Batten down the hatches!" when the going gets rough. As a writer, you are the captain of your ideas. You're in control and must use all your skills to make sure no obstacles prevent your message from getting through. In this unit you've revised several pieces to increase the fluency and practiced using strategies to make even rough drafts stronger. Now it's time to put these skills together. Ready? Then smooth sailing ahead!

Creating a Draft

Time to create a rough draft. This portion of the lesson has two parts, and we think you'll really like the first part.

Part 1

Coach your teacher in choosing a topic he or she could write about spontaneously as you create your own rough drafts. Choose a topic your teacher knows well enough to write about without research and likes well enough to write about with strong voice. Each writing circle has five minutes to think of one or two suggestions, no more. Choose wisely. When you offer your suggestions, your teacher will make a choice but may or may not tell you what it is (just yet).

Sentence Fluency

Name _____ Date _____

> **Part 2**

For this second part of the lesson, you need to choose a topic for yourself. Choose any topic you wish, or use our list to help you think of an idea.

★ My topic _____
- Secrets to writing with fluency
- The most fluent writer (in my humble opinion) on the planet
- The best grilled chicken (vegetables/steaks/fish) ever
- Guidelines for effective blogging
- Technology in twenty years

Prewriting

Once you have your topic clearly in mind, spend about five minutes prewriting. You may choose to complete any of the following.

- Make a sketch
- List details
- List questions a reader might have

Or do anything else that gets the wheels turning and the thoughts flowing.

Drafting

Use the remaining class time to write. If you get stuck, do one of the following things.

- Look back at your prewriting for a new idea.
- Read the model passage you identified as your inspiration.
- Try to write just ONE more sentence (then one more after that).
- Keep the writing flowing as much as possible. Stay quiet so all the writers (including your teacher) can concentrate.

Sharing and Plotting Your Revision

Find the draft you completed in the first part of this lesson. Read it over quickly and quietly to yourself, making any last-minute changes you wish. Then share your writing aloud in your writing circle. As you share or listen, keep the following suggestions in mind.

As a writer . . .

Don't apologize for your writing. Just read. Keep your mind open. Make notes directly on your work about helpful suggestions. Use any helpful comments and insights when you revise. Remember, you can't expect others to hear your writing exactly the way you do. That's why you need another perspective!

As a responder . . .

Focus your comments on fluency. Ask questions and make positive suggestions about strategies to improve sentence fluency. Remember, all comments should be supportive.

Charting a Course for Revision

On your own, plan your revision. You should have ideas from many sources, such as the following.

- Your group's comments
- Your class list of fluency secrets
- Your own notes
- Your model passage and any other literature that inspires you

Sentence Fluency

In addition, we're offering you a quick checklist of things to try. Read the following and check at least two things you feel you'll do as you revise.

- [ ] Vary sentence beginnings
- [ ] Vary sentence lengths
- [ ] Connect ideas logically
- [ ] Read aloud to check the flow of ideas
- [ ] Combine short choppy sentences to make one smooth sentence
- [ ] Divide overly-long sentences into two or more smaller sentences
- [ ] Correct run-ons
- [ ] Use repetition for special effect
- [ ] Try writing a sentence several ways to see what works best

Use scratch paper for your revision. Write for at least 15 minutes. Your teacher will let you know when three wrap-up minutes remain.

Share and Compare

When you finish revising, meet with a partner and take turns reading your revised work aloud. Discuss the kinds of changes you made and the strategies you used. Give your partner specific, positive feedback—not just suggestions, but what you liked. Take in your partner's comments about your work, as well. After sharing, take another 3–4 minutes to make any final changes you thought of during your sharing time.

Sentence Fluency

Name _____ Date _____

Revision in Action

Your teacher is about to share his or her rough draft and revision. As you watch and listen, be looking for specific strategies your teacher used to improve the fluency of the rough draft. Comment on what you like or might try in your own work. If you hear or see something (however small) that should be added to your list of fluency secrets, be sure to mention it.

A Writer's Questions

As you plan your revision, it's helpful to have a partner with whom to share your writing. What can you do if a partner is not available? Can you coach yourself?

Putting It to the Test

Even though you cannot really study for an on-demand writing test, are there things you can do to prepare? What are some things a writer could do the night before on-demand writing to help make fluency (or anything about the writing) stronger?

Conventions and Presentation
Editing Level 1: Conventions

Discriminating Fragments

Many writing teachers feel it's OK to occasionally bend the rules of formal grammar and usage in writing, but only after learning what those rules are. Why are they so picky about that part? They want to make sure that when you bend those rules, you'll be doing it intentionally and with a clear purpose in mind. Also, to be effective, rule bending has to come in small doses. For example, an occasional fragment can be powerful. A writer who writes primarily in fragments may distract and confuse readers.

In his classic survival story *Hatchet,* Gary Paulsen uses fragments frequently. One very powerful moment comes at the end of the first chapter. A small plane in which the hero, Brian, was riding has crashed in the remote wilderness and the pilot has had a heart attack. One word sums up Brian's thoughts and feelings at this moment: "Alone." This single word is an effective fragment, but Paulsen also uses it as the entire final paragraph of that chapter. A one-word paragraph? Most unusual. Why might that work in this instance?

Alone.

A Warm-Up: Part 1

To figure out when or why you should use fragments, you need to be able to tell the difference between a fragment and a complete sentence. So let's begin with some definitions. On the following page, write down your first thoughts without looking at any reference books.

1. A complete sentence is

_____

_____

2. A fragment is

_____

_____

Now use your personal definitions to help decide whether each item below is a complete sentence or a sentence fragment. Put an FR (for fragment) in the blank to mark each group of words that does not form a complete thought.

HINT: Length is not always a good indicator!

___ 1. As I was crossing a busy street carrying two heavy bags of groceries and trying to hold my little sister's hand.

___ 2. I don't know how some people manage to get going in the morning without eating breakfast.

___ 3. Dogs ran.

___ 4. The plan.

___ 5. Anger!

___ 6. Run!

___ 7. Good grief!

___ 8. Running for cover.

___ 9. Keep safe.

___ 10. On the other hand, if we have time after dinner.

Share and Compare

Meet with a partner to share and compare your definitions and then your responses to each of the exercises. Do you and your partner agree on your definitions? Did you identify the same fragments from our list of possibilities? Complete the following steps.

1. Turn any two fragments from the preceding list into complete sentences.

2. Take time to clarify your definitions of *fragment* and *complete sentence,* using any resources available in your classroom.

A Warm-Up: Part 2

In each of the next two writing examples, the writer has used sentence fragments. Read each example aloud carefully, and underline any fragments you notice. Then decide if you feel the use of fragments is effective or just distracting.

Example 1

Winter is a long season where I live. Snow, ice, cold. More snow. Then more ice and cold. It all begins in late October and doesn't let up until well into April. I don't like it, but I've learned to live with it.

☐ This was a very effective use of fragments.
☐ I found these fragments just plain distracting.

Example 2

> Recently. A small gelato shop. Opened in my neighborhood. In case you don't know, gelato is Italian. Ice cream. They have so many. Great flavors. Hazelnut. Raspberry. Lemon. Pistachio. Really delicious. Try them. All.

☐ This was a very effective use of fragments.

☐ I found these fragments just plain distracting.

Making Fragments Rock

Now it's your turn. Choose any topic (including one from our list). Then write a short piece in which you use just one or two fragments on purpose. No accidental fragments, please! Make them effective. Use them to draw attention to a particular word, phrase, or idea. Remember, effective fragments do the following things.

- Make sense because they gain meaning from the surrounding sentences
- Emphasize a particular word or idea
- Sometimes answer a question raised by the writer (What did I need more than anything? Food.)

Possible topics:

★ My topic _____

- A pesky brother, sister, cousin, pet
- A food you can keep
- Frustration with _____ (you name it)
- A tough habit to break
- Why I do (or don't do) well on tests
- The best person I know
- I don't know why it bothers me, but . . .

Prewrite for 3–5 minutes. Draw, list details or questions, or read a passage that makes good use of fragments. Then write your draft on scratch paper. Underline your fragment(s).

Share and Compare

Share your writing with a partner or in a writing circle. As you listen, see if your ear picks up a fragment the writer is using. After that writer finishes, share what you heard. See if it matches what the writer underlined. Talk about fragments you found especially effective.

A Writer's Questions

As a writer, do you have to be careful about when or where you break the no-fragments rule? Why? What do you think when you run across fragments in the writing you read? Do they bother you, or do you think they sometimes add to voice?

Editing Level 2: Presentation
No Rules Poetry

In Sharon Creech's *Love that Dog,* the main character, Jack, is frustrated with being asked to write poetry. He claims that poetry is really just about making short lines. Do you agree or is there more to it?

As you'll see in this lesson, lines of poetry can be long or short, and they can comprise whole sentences, fragments, phrases, or single words. Sometimes the lines rhyme, sometimes not. The rhythm may be obvious or subtle. And that's not all. Sentences within poems may span many lines or just one line each. What's more, poems can follow the usual conventions of punctuation, grammar, capitalization, and spelling, or they can break all the rules. They come in countless shapes, as well. Behind every decision is a poet trying to get a message across in a way readers will remember, and perhaps return to many times.

A Warm-Up

Your teacher has gathered a variety of poems for you to explore and read. Spend some time looking first, then choose as many as you think you can share aloud in your writing circle within about 10–12 minutes.

As you make your choices, notice the following about each poem.

- Length
- Format
- Words per line
- Alignment (left side of page, centered, shaped, etc.)
- Use of single words, phrases or fragments, whole sentences
- Use of rhyme
- Rhythm and readability

Don't worry too much about message right away. Sometimes that doesn't strike you until you have read a poem several times. Poems should be read again and again, perhaps at different times in your life. As you and others share aloud, record favorite titles and poets' names on the following list.

Poem Title **Poet's Name**

1.

2.

3.

4.

5.

6.

7.

8.

(Use your own paper if you need to make a longer list.)

Choose one poem that stood out from the others. Maybe you were caught up in the message or you just liked the sound of it. Share your thoughts about it here.

My thoughts about _____ (Title)

_____

_____

_____

Share and Compare

Make a class list of some of the poems that interest you. Is there any one poet whose work you'd like to explore further?

The Finishing Touch

Here are some starts (just a few lines) to three poems, each with a very different style. Read each partial poem aloud, carefully, in your writing circle. Listen for ideas that stay in your heads—words, phrases, or images that speak to you or capture your imagination. Choose one poem to finish, and take about five minutes to add as many lines as you think of. Give your poem a title.

Poem 1

>Sometimes
>when your head
>and heart
>are full of too many people
>shouting
>too many cars
>rushing
>too many doors
>slamming

Poem 2

>On the front wall of my math classroom
>Next to a picture of Pythagoras
>My teacher had written
>   "Life is hard—write it down."
>On the back wall of my English classroom
>Next to a photograph of Dr. Seuss
>My teacher had written

Poem 3

>I have collected things
>for as long as I can remember–
>Age 1–family faces
>Age 4–sticks and frogs
>Age 5–friends
>Age 6–shells and rocks
>Age 7–teachers' comments
>Age 8–favorite games
>Age 9–favorite books

Share and Compare

Be ready to share your poem with the whole class. Take turns reading your poems aloud, from the beginning, not just from where you started writing. How difficult was it to fit into the rhythm and thinking of another poet?

A Writer's Questions

What is your experience with poetry? In your opinion, is writing poetry harder than writing an essay or story? Is writing poetry just about making short lines?

Presentation Matters

Whether you like, love, tolerate, or really dislike poetry, you probably have an opinion about what makes a good poem. Here's a chance to put your ideas together in writing a poem of your own. Follow these six steps:

1. Take out the small object you brought with you to kick off your writing today.

2. Hold it in your hand, turning it over and over to really appreciate its surface features. Look at it closely, noticing small details you may never have noticed before.

3. If you have a jeweler's loupe (small magnifying glass), use it to notice even more details about this object.

4. Ask yourself, "What does this make me think of?" The memories or associations it calls up may surprise you.

5. Make some notes on scratch paper. Write whatever comes into your head—as fast as you think of it. Don't shut anything out. Keep this up for 2–3 minutes.

6. Then begin your poem. Make it as long or short as you wish. Write for about ten minutes. Remember, the poem doesn't have to be about the object itself, though it can be. It could also be about any memories or experiences that object calls to mind.

7. As time permits, share your poem with a partner or in a writing circle.

Sample Paper 19

Score for Sentence Fluency _____

Mosquitoes, Beware!

A new kind of mosquito repellent is being developed. It's being developed by scientists right now. This is really good news for campers. This is good news for ALL outdoor people! This new repellent is made from tomatoes. It is hard to believe, but it is true. It's made from tomatoes. Yes! This is true! Scientists found out that mosquitoes hate tomatoes. There's this natural compound in tomatoes. This natural compound actually protects them from insects. Scientists used this natural compound to make a mosquito repellent.

The new repellent really works. It really does! It works even better than you might think! It repels more than mosquitoes. It repels ticks, ants, biting flies, fleas, cockroaches, and aphids, too. The good news for people is that it is safe. The tomato repellent is safer than mosquito repellents made from other chemicals. The tomato repellent has no bad side effects. The chemical repellents, moreover, can have many side effects. Chemical repellents can cause skin rashes, swelling, itching, and other irritations. Just to name a few. The tomato repellent doesn't cause any of these things. It just keeps mosquitoes away.

Mosquitoes are terrible pests that carry diseases. If we have a harmless way to get rid of mosquitoes, we should use it. If this kind of tomato repellent works, maybe scientists will develop other products that use tomatoes. If you are thinking you will just rub tomatoes on your skin, forget it. It won't work. It won't hurt you, but it won't work. It could save you money, of course! (It won't work, though, seriously.) Mosquitoes, beware!

Sources

Jacobson, Cliff. 1987. *Camping's Top Secrets: A Lexicon of Camping Tips Only the Experts Know.* Guilford, CT: Falcon Guide.

Cliff DiConsiglio. *Blood Suckers!: Deadly Mosquito Bites (24/7:Science Behind the Scenes).* New York, NY: Scholastic, 2008.

Kalman, Bobbie. **The Life Cycle of a Mosquito.** Published in New York, NY by Crabtree Publishing in 2004.

Rooney, Darlennia. "Organic Mosquito Repellent." *Associated Content.* 14 April 2009 <http://www.associatedcontent.com/article/1636247/organic_mosquito_repellent.html?cat=69>

"Mosquitoes Repelled by Tomato-Based Substance; Safer, More Effective Than DEET." *Science Daily.* 11 June 2002 <http://www.sciencedaily.com/releases/2002/06/020611070622.htm>

Sentence Fluency

Name _____   Date _____

Sample Paper 20
Score for Sentence Fluency _____

Starlings: Cheep! Imitators

"The phone's ringing," my father said, never looking up from his workout on our deck. The phone rang again, and this time I heard it, followed by my father's slightly more insistent voice. "Hey, somebody get the phone! Someone's cell phone is ringing."

"Dad, the phone isn't ringing," I replied as I continued making myself a sandwich. Trust me, if there were a phone ringing somewhere in our house, I'd be answering it. I could see him through the kitchen window, and he looked up, just as we heard another ring. We stared at each other, wondering what to make of the sound. Clearly, someone's phone was ringing, but it wasn't ours. My dad's cell phone was on the deck railing, and mine was in my jeans pocket. Our house phone was quiet on the kitchen counter. No one else was around—anywhere. The phantom phone rang again. "Well, what in blazes . . .," my dad started to ask.

Just then a starling flew from a branch just over Dad's head. As crazy as it sounds, the bird was "ringing" as it flew away. "Dad," I laughed, "I think it's for you!" Even he had to laugh.

We discovered what scientists have known for years—that starlings, and their close relatives, grackles, can mimic almost any kind of sound, including the ring tone of a cell phone. In the wild, these birds are more likely to imitate the calls of other birds in their territories. But in captivity, starlings have been known to imitate human speech, music, bells, and various sounds from television or radio. They have even been trained, like a parrot or myna, to converse with their trainers.

For many people, starlings are those annoying birds who gather in trees or on houses during winter in such large numbers that they can actually do damage as they scratch for food and leave behind a whitewash of droppings. They're notoriously unpopular in the Midwest, where they descend in virtual black clouds on crops, consuming enough grain to put a serious dent in farmers' budgets. But before they are written off entirely as pests, people should consider their amazing imitative abilities. The next time you wonder whose cell phone is ringing, check the nearest tree branch. It could be a starling, one of nature's annoying yet extraordinary wonders.

Sources

Bronson, Wilfrid S. *Starlings.* Santa Fe, NM: Sunstone Press, 2009.

Roth, Sally. *Backyard Bird Secrets for Every Season: Attract a Variety of Nesting, Feeding, and Singing Birds Year-Round.* New York, NY: Rodale, 2008.

Wagner, Jack. "Pest Bird Species." *Bird busters.* 1985 <http://www.birdbusters.com/bird_control_starling.html>

Sentence Fluency

Name _____ Date _____

Revising Checklist for Sentence Fluency

☐ I read this aloud. It's smooth and easy on the ear. The writing really *flows!*

☐ I <u>underlined</u> sentence beginnings (first three to four words) to check for variety.

☐ MANY sentences begin in different ways with words that connect ideas. OR

☐ I highlighted beginnings that could use revision.

☐ Some of my sentences are long and flowing—combining several ideas. Others are short and snappy.

☐ I checked for sentence problems and (as needed) did the following to revise.

　☐ Combined some choppy sentences to make one smooth sentence

　☐ Got rid of run-ons

　☐ Got rid of fragments I did not *mean* to write

　☐ Rewrote sentences that did not sound as fluent as I wanted them to

☐ IF I used dialogue, I read it aloud to make sure it sounded like real conversation.

☐ _____ rated my writing:

| 1 | 2 | 3 | 4 | 5 | 6 |

☐ I used punctuation (and perhaps *italics* or ALL CAPS) to make sure readers would read my writing with *just* the right inflection.

> **Note** When it comes to checking fluency, *nothing* takes the place of reading aloud. Remember, just because a sentence is grammatically correct, that's no sign it's fluent and beautiful. You may need to write a sentence three or four ways to discover what makes it sing. Did you do that?

Student Rubric for Ideas

6
- My main message or story is clear and will hold your attention.
- I know this topic inside and out and take readers on a journey of discovery.
- I included intriguing details a reader will notice and remember.
- My writing makes a point—or focuses on a clearly defined message or issue.

5
- My main message or story is interesting and easy to understand.
- I share important information—and tell enough to give readers a full picture.
- My paper contains many interesting details.
- I narrowed my topic enough to give readers an in-depth look at my subject.

4
- A reader can identify my main idea or make sense of my story.
- I have enough information for a first draft, but more would help.
- My writing includes a few interesting details. Readers might want more.
- I think I need to narrow my topic a little. I'm trying to cover too much.

3
- A reader can guess what my main idea is—or tell what my story is about.
- I know enough to start—then I have to make things up.
- My details are general—things many readers already know.
- My topic feels way too BIG. I can't cover everything.

2
- A reader might have trouble figuring out the main message here.
- The story or message isn't really clear in my mind. I just wrote to fill the page.
- I repeat things—or stop when I run out of things to say.
- I bounce from topic to topic—or list thoughts at random.

1
- I put my first thoughts on paper. You couldn't call it an essay or story—yet!
- I'm still figuring out my topic.

Student Rubric for Conventions and Presentation

6
- A reader will have to look hard to find errors in my writing!
- I edited carefully, reading silently and aloud. This is ready to publish.
- I used conventions to bring out the meaning and voice.
- My presentation has eye appeal and makes information easy to find.

5
- A careful reader might find minor errors—but nothing serious.
- It *might* need a few touchups, but it's *almost* ready to publish.
- My conventions support meaning and voice.
- My presentation makes important information stand out.

4
- Errors are noticeable, but they won't slow readers down.
- I need to go over this one more time, reading aloud as I edit.
- My conventions support the message and make reading fairly easy.
- My presentation is OK—it draws attention to key points.

3
- Readers might notice the errors more than the message!
- This writing needs *a lot* of editing.
- Mistakes could puzzle readers or force them to read some things twice.
- I need to work on presentation. Readers can't tell what to focus on.

2
- Parts of this are not edited at all. Mistakes jump right out!
- I need to go over this line-by-line, pencil in hand, reading aloud.
- Readers will need to "edit" as they read—that should be *my* job!
- I did not think about presentation yet.

1
- Mistakes make this hard to read, even for me.
- I have not done any editing yet—I'm not sure how to begin.
- Even if they read it two times, I'm not sure readers will get the message.
- I need help with editing and presentation.

Student Rubric

Student Rubric for Organization

6
- Everything connects in some way to my MAIN message or story line.
- My paper is easy to follow, even with a quick reading. It has some twists and turns, though—to make reading interesting!
- The lead is striking and will pull readers right in.
- The conclusion is original. I want to leave my readers thinking.

5
- I stay focused on the discussion or story all the way through.
- You can easily follow my "trail of thought."
- You'll like my lead, and it will hook you.
- My conclusion is satisfying. It wraps up the discussion or story.

4
- I may wander here and there, but I think you'll see a connection.
- I think you can follow this pretty easily.
- My lead sets things up. It kicks off the story or discussion.
- My ending wraps things up.

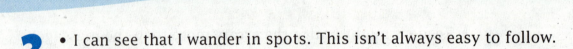

3
- I can see that I wander in spots. This isn't always easy to follow.
- I need more surprises! Readers can tell exactly what is coming!
- I have a lead. It could be more exciting.
- I have a conclusion. It's probably one you have heard before.

2
- I jump from topic to topic. This is really hard to follow, even for me.
- This writing is like a messy closet! I need to move some things, or toss some out.
- I need a new lead.
- I need a new conclusion, too.

1
- I just wrote to get something on paper.
- Nothing goes with anything else. Don't look for a pattern!
- I didn't know how to begin.
- I didn't know how or when to stop, either.

Student Rubric for Voice

6
- This is ME. You can hear my voice in every line.
- A reader would *love* sharing this aloud.
- This topic matters deeply to me—I said exactly what I felt and thought.
- I wanted to reach readers—to make them feel the way I feel.

5
- This voice sounds like me—it doesn't blend in with others.
- I think some readers would share this writing aloud.
- Reading this writing will convince you I care about my topic.
- This voice fits my purpose—and will get readers involved.

4
- I think my voice stands out from many others.
- There are some good moments to share.
- I care about this topic. I think that comes through in many parts.
- I think my voice will speak to many readers.

3
- I hear my voice in *parts* of this.
- With a little work, parts would be ready to share.
- I tried to sound excited—I couldn't do it all the time.
- This voice won't reach all readers.

2
- This voice blends with many others. There's barely a whisper of ME.
- I don't feel ready to share this writing—yet.
- I need a topic I know more about—and care more about.
- I'm still figuring out my purpose and who my readers are.

1
- There is nothing in this writing to make it mine.
- There's no reason to share this aloud.
- I don't have any strong feelings about this topic.
- I wrote what I had to write to finish the assignment.

© Great Source. Permission is granted to copy this page.

Student Rubric 233

Student Rubric for Word Choice

6
- I chose the right word(s) for the moment—many are memorable enough to highlight.
- Strong verbs create vivid, moving pictures in the reader's mind.
- Carefully chosen sensory details pull readers right into my experience.
- You won't find clutter. Every word counts!

5
- My words are clear. I found *my own way* to say things.
- I used many strong verbs to add life and energy.
- Sensory details make my descriptions vivid.
- All words add to the message. I wouldn't cut anything.

4
- My words are used correctly.
- I used *some* strong verbs—as well as descriptive words.
- I used sensory details—if they fit. And I didn't overdo it.
- Wordiness and repetition are not problems.

3
- I used too many general words: *nice, great, cool, neat, good.*
- If you look hard, you might find a strong verb hiding in there.
- I need more sensory details. OR . . . I used TOO many!
- Parts are wordy—or I repeated certain words too many times.

2
- I didn't really reach for the right words—or the best words.
- I didn't use any ACTION verbs. I don't have much description either.
- I didn't think about *sounds, feelings, smells,* or *tastes.*
- Repetition is a problem. Sometimes I went on and on without saying anything new.

1
- I wrote the first words that came into my head.
- Getting *any* words on paper was hard for me.
- I'm not sure readers can picture anything or even tell what I am trying to say.
- I need help finding words that will let me say what I mean.

Student Rubric for Sentence Fluency

6
- My writing is smooth and easy to read on the first try.
- Sentences differ in length and begin in ways that show how ideas connect.
- You can read this expressively to bring out every ounce of voice.
- If I used dialogue, it's so real you can perform it like a play.

5
- My writing flows smoothly. No bumps or sudden stops!
- Sentences differ in length and structure.
- It's easy to make this paper sound fluent and smooth.
- If I used dialogue, it sounds like real conversation.

4
- My writing is easy to read with a little practice.
- There's enough variety to make sentences interesting.
- With a little effort, you can make this writing sound fairly fluent.
- If I used dialogue, it's pretty realistic.

3
- Some parts are smooth—others are choppy or rambling.
- Too many sentences start the same way or are the same length.
- To read this smoothly, you need to rehearse and pay attention.
- If I used dialogue, it needs work.

2
- Choppy sentences, run-ons, or other problems make reading slow.
- I use the same sentence patterns over and over. I might have fragments I didn't want.
- If reading this aloud, prepare to hit some bumpy spots.
- I didn't use dialogue. Or I couldn't make it sound real.

1
- I'm not sure all my sentences are really sentences.
- It's hard to tell where my sentences begin and end.
- Readers need to fill in missing words or punctuation.
- I didn't try to write dialogue.

Student Rubric